RESTRUCTURING REGIONAL AND LOCAL ECONOMIES

Restructuring Regional and Local Economies:

Towards a Comparative Study of Scotland and Upper Silesia

Edited by
GEORGE BLAZYCA
University of Paisley UK

Routledge
Taylor & Francis Group

LONDON AND NEW YORK

First published 2003 by Ashgate Publishing

Reissued 2018 by Routledge
2 Park Square, Milton Park, Abingdon, Oxon OX14 4RN
711 Third Avenue, New York, NY 10017, USA

Routledge is an imprint of the Taylor & Francis Group, an informa business

A Library of Congress record exists under LC control number: 2002111306

ISBN 13: 978-1-138-71863-0 (hbk)
ISBN 13: 978-1-138-71861-6 (pbk)
ISBN 13: 978-1-315-19444-8 (ebk)

Contents

List of Figures

List of Tables

List of Contributors

George Blazyca is Professor of Economics and European Studies in the Paisley University Business School and co-ordinator of the University-wide Centre for Contemporary European Studies.

Chik Collins is Lecturer in Politics in the School of Social Sciences, University of Paisley, Scotland.

Małgorzata Czornik is a lecturer and researcher in local economic development in the Department of Strategic and Regional Sciences *(Katedra Badań Strategicznych i Regionalnych)* at the Academy of Economics, Katowice, Poland.

Mike Danson, is Professor of Scottish and Regional Economics in the Paisley University Business School, a member of the University's Centre for Contemporary European Studies and chair of the UK-wide Regional Studies Association.

Adam Drobniak is a researcher, specialising in FDI in regional development in the Department of Strategic and Regional Sciences at the Academy of Economics, Katowice.

John Foster is Professor of Politics and Sociology and Head of the School of Social Sciences at the University of Paisley.

Krystian Heffner is Professor of Local Economic Studies in the Department of Strategic and Regional Sciences at the Academy of Economics, Katowice.

Janusz Hryniewicz is deputy director of the Institute for Regional and Local Development *(Instytut Rozwoju Regionalnego i Lokalnego)* at Warsaw University.

Ewa Helińska-Hughes is Research Fellow in the Centre for Contemporary European Studies, University of Paisley.

Andrzej Klasik is Professor of Regional Economic Studies and Head of the Department of Strategic and Regional Sciences at the Academy of Economics, Katowice.

Marek S. Szczepański is Professor of Sociology at the Silesian University in Katowice, Poland.

Bogumił Szczupak teaches and researches in the Department of Strategic and Regional Sciences at the Academy of Economics, Katowice.

Geoff Whittam is Senior Lecturer in Economics and Enterprise in the University of Paisley Business School.

Krzysztof Wrana teaches and researches in the Department of Strategic and Regional Sciences at the Academy of Economics, Katowice.

Alistair Young is Professor Emeritus of Economics, University of Paisley.

Acknowledgements

The material gathered in this volume grew out of a joint Katowice-Paisley seminar held at the University of Paisley in October 2001. I am very grateful to all who took the time to prepare papers for that meeting. I am even more grateful that authors worked hard and uncomplainingly to revise papers for publication to a tight deadline.

Our seminar and the volume that results benefited enormously also from the input of discussants and chairpeople including Heather Sim (Scottish Enterprise), Jim Cunningham (Head of Economic Development, Renfrewshire Council), Bob Arnot (Glasgow Caledonian University), David Deakins from the Paisley Enterprise Research Centre, Martin Myant and John Struthers from Paisley's Europe Centre, none of whom, it goes without saying, are responsible for what is written in what follows.

Without the financial support of Paisley's Centre for Contemporary European Studies and the Paisley Business School, our 2001 seminar would have been impossible and that support is acknowledged here.

As editor, my job was made much easier by the supportive and patient responses of all contributors to the many 'small points' that I found myself generating as I assembled pieces for publication. For that, many thanks to all. Both I and Andrzej Klasik are grateful also to Mike Danson, Chik Collins, John Foster and Krystian Heffner for comments on our introductory 'overview'. Ewa Helińska-Hughes, from the Centre for Contemporary European Studies, helped smooth Paisley – Katowice communications and even served as unofficial simultaneous translator during some seminar sessions. Doris Cairns of the Paisley Business School office worked extremely hard in preparing the camera-ready copy, displaying huge initiative and spending time beyond the call of duty on tables, diagrams and maps. Her 'Don't panic' assurances were also much appreciated.

Academics always owe a lot to the other life support mechanisms that seem to kick into place at busy times, so last, but by no means least, my very special thanks to Wendy Blazyca for her tolerant understanding, especially when she heard the ominous phrase 'I'm just off to do some editing'.

George Blazyca
Paisley
February 2003

PART I

INTRODUCTORY OVERVIEW

Chapter 1

Transforming (worn out) Industrial Dynamos into Strong Regional Economies – A Comparison of the West of Scotland and Upper Silesia

George Blazyca and Andrzej Klasik

Introduction

Post communist transformation and the restructuring of the heavy industrial basin of Upper Silesia in Poland may, at first sight, seem remote from the problems of regional development in Scotland. Yet parallels exist. At one time the West of Scotland with its concentration of basic heavy industry was the 'Upper Silesia' of Great Britain, Katowice had something in common with Glasgow, just as Łódź was the counterpart of Manchester in cotton and textiles.

Of course, there are very many differences too and it is not our intention to overload the Upper Silesia/West of Scotland comparison. The first thing to notice is time displacement. Upper Silesia is trying to find a way to deal with heavy industry restructuring today while in Scotland that struggle ended in 1993 when the giant Ravenscraig steel works in Motherwell finally closed. There are also huge differences of a 'systems' nature. Heavy industry restructuring is taking place in Upper Silesia alongside the complex business of post-communist economic and social transformation. Yet despite the differences, restructuring heavy industry is a strong enough common denominator to permit the question, does the Scottish experience have anything to offer the people of Upper Silesia, the *Ślązacy*?

As we probed diverse experiences of industrial restructuring, as well as economic and social aspects of regional development, the similarities and the differences between our two cases became clearer. In the introductory notes that follow we set the scene, drawing attention to some lessons and touching on some ideas for future research.

The Industrial Dynamos Wear Out

The West of Scotland

Around 100 years ago, the West of Scotland had 'the biggest concentration of heavy industry in Britain' (Foster, 2001, p.417). Coal and steel supported heavy engineering, the products of which, especially the locomotives and ships, found ready markets through the empire and elsewhere. Although not immediately apparent, the First World War changed everything. From the 1920s Scottish heavy industry moved into a long term decline punctuated only by momentary relief when short periods of boom masked underlying forces. In the 1950s, with competitors for a short time weakened and in disarray, good times returned. In those days Scottish shipbuilding and heavy engineering employed almost 100,000 and mining another 80,000, virtually the same as in 1914. It was the high point of heavy industry in Scotland and coal, steel, iron and engineering accounted for around 25 per cent of the insured labour force (Devine, 1999, p.548). In 1955 shipbuilding on the River Clyde reached a post war peak with 485,000 gross registered tons launched, one third of the British total (Johnman and Johnston, 2001, p.121).

In the years that followed, the decline of heavy industry in the West of Scotland was almost relentless although by no means steady. The de-industrialization of the UK economy that commanded so much academic and policy-making attention in the 1970s was soon to shift up several gears as 'Thatcherism' kicked in. Mrs Thatcher's particular blend of authoritarian politics, masked in the liberal language of individualism with its supposedly greater 'choice', and firmly embedded in a Chicago-style free-market economics had immense and immediate consequences for the UK and Scottish economies. Its influence soon spread. In Poland Mrs Thatcher, the 'Iron Lady', was much admired despite the fact that by the early 1990s the negative consequences in the UK of the Thatcher 'experiment' were becoming apparent.

Scotland experienced a Thatcher induced 'shock-therapy' in 1979-81 when manufacturing output fell by 11 per cent and employment by 20 per cent. Perhaps modest by the standard of later shocks elsewhere – especially in Poland over 1989-91 when measured output (GDP) fell by 18 per cent and manufacturing production by 33.4 per cent – it was nevertheless painful.[1] It set the scene for the last acts in Scottish de-industrialization. In the longer interval from 1976 through to 1987 Scotland lost almost 30.8 per cent of its capacity in manufacturing and 36.9 per cent in heavy industry. During the 1980s the number of working coalmines fell from 15 to only 2. Employment in manufacturing fell from 800,000 in the 1980s to less than half that number in the 1990s. Shipbuilding, the jewel in the crown of Scottish heavy industry, had almost entirely disappeared by the start of the new millennium. As the leading historians of Scottish shipbuilding put it 'a once great industry [with 100,000 employees in 1921] had become little more than a rump employing a few thousand people' (Johnman and Johnston, 2001, p.126).

---

[1] Data on Poland from *Rocznik Statystyczny*, (1994), p.309. Data on Scotland from Devine (1999), p.592.

The concentration of heavy industry in the West of Scotland meant, as in post-communist Upper Silesia, that a sectoral problem was immediately also a regional problem. It was a problem easy to diagnose. In 1961 the then Chairman of Ferranti, Sir John Toothill, pinpointed, in an official report, 'the major structural weakness of the Scottish economy as the over reliance on traditional heavy industry, the inability to adapt to world markets, and the failure of the new science-based manufacturers to become established on any significant scale' (Devine, p.572).

The reasons for the failure to adapt still arouse controversy. For some the decline was the inevitable consequence of globalization. For others the problem was more directly political. John Foster argues in this volume (see chapter 5) that from the 1950s the government adopted policies that reflected the interests of the large City of London-owned transnational companies. These policies sought to secure fast capital concentration and accumulation at British level. The Toothill Report, far from being disinterested, represented, in John Foster's view, a legitimization of the process. Its categorization of production as either 'traditional' or 'new and science based' consolidated attitudes that blocked the way for comprehensive and integrated development. The significance of shipbuilding as a relatively hi-tech end-user of heavy industry products was ignored and instead supplies of skilled labour and scarce materials such as steel were switched to the incoming branch plants of the British transnationals. Restructuring, he argues, was driven by an 'external agenda', addressing needs that 'were not those of Clydeside but of British business at UK level'.

In the 1960s UK governments sought to deal with the problem of the West of Scotland by inducing or leaning on businesses to relocate from other parts of the UK. Those policy initiatives, though undoubtedly bold – taking a giant steel strip mill to Motherwell and volume car assembly to Linwood, near Paisley – proved ultimately ineffective. Meanwhile, shipbuilding, that critically important sector in heavy engineering, continued to lose competitiveness.

From the mid 1970s, the regional policy focus switched to inward investment as the key to future prosperity. The *Scottish Development Agency* (SDA), and later *Locate in Scotland* (LiS), joined the fierce international struggle to entice multinational capital to set up in Scotland. For a time that strategy seemed to work. 'Silicon Glen' – the electronics/IT investment along the Scottish central belt (roughly along the M8 motorway between Glasgow and Edinburgh), Scotland's modest and less sunny equivalent to California's 'Silicon Valley' – looked to be its chief success, an experience carefully assessed by Alistair Young in chapter 12 of this book. As heavy industry shed labour at an accelerating rate in the 1980s the opportunities for entrepreneurship, at least in the Thatcher model of economic development, should have been on the increase. A great propaganda push in favour of entrepreneurship was launched. The public agencies were rebranded as '*Enterprise* this or that' and the lead development body, *Scottish Enterprise*, (a transformed SDA), gave itself the task, among others, of stimulating the 'Business Birth Rate'. The switch in strategies did have an impact on the economy, if not on the business birth rate, as much of the central belt of Scotland became a manufacturing-assembly base for multinational firms as well as a giant services centre. For some this was a useful structural diversification but others had doubts on how robust the new economy would be.

This push in favour of entrepreneurship was also reflected in urban policy. In the late 1980s the government believed that the projection of an 'enterprise culture' into some of Scotland's poorest and most stigmatized localities would see them transformed into stable and thriving communities. In his evaluation of this *New Life for Urban Scotland* (Scottish Office, 1988) initiative in chapter six below, Chik Collins argues that these hopes were to prove illusory. Here, perhaps, is one part of the Scottish experience which others will not want to replicate.

Over the years European Union aid was eagerly sought by regional authorities becoming increasingly adept at lobbying for designation and winkling the funds from Brussels. The latest turn in regional policy (reviewed more fully by Mike Danson in the chapter that follows), such as it is in the Europe of the single market, is that attracting inward investment, important though it is, may be less critical to the long term good health of regional economies than developing a highly skilled labour base, promoting the development of human capital and the other elements that contribute to the highly fashionable 'post neo-classical endogenous growth theory'. In addition, the view is developing that stronger regional economies are those that are able to support industry 'clusters' hi-tech or otherwise.

Upper Silesia

The huge coal and complex ore deposits of Upper Silesia are the foundation of its development as a heavy industry region. After coal came steel and the economic character of the region was firmly established. It was a region that shared with the West of Scotland the problems associated with industrial monoculture although perhaps more severely since Upper Silesia had little of the technically advanced heavy engineering associated in Scotland with shipbuilding. In the first half of the twentieth century Upper Silesian industry was buffeted by the same international economic gales that hit the West of Scotland. The impact of international political developments was however very different. Upper Silesia was a border region in the pre-1914 system of European empires. From the pre-industrial period until the First World War, it was part of the German Empire, its principal town, *Kattowitz*, became the Polish *Katowice* only in 1922. But Upper Silesia remained split between its Polish and German parts complicating industrial development through the interwar period. From 1939 to 1945 the region was incorporated once again into the German Reich, reverting to Poland in the post 1945 division of Europe. As might be expected, until 1945, German capital was primarily responsible for the region's industrial development and its main role was to supply German industry with raw materials.[2] After 1945 it continued to fulfil the role of raw material supplier, either to support the post-war drive for domestic (socialist) industrialization or simply for export, to both Comecon and Western markets.

---

[2] Landau and Tomaszewski remind us that the 'products of the Silesian heavy industry were mainly sold in Germany and much less frequently exported to the Russian or Austrian partition' (Landau and Tomaszewski, 1985, p.14), moreover, 'In Upper Silesia mining developed above all on the large estates belonging to the German aristocracy: there was no room for Polish capital' (Ibid, p.18).

The region's mono-structure not only remained but also intensified and was invested with the 'heroic' qualities of socialist labour. Miner (*górnik*) and steelworker (*hutnik*) were the envied advance guard of the new working class. The drive for output was the key goal of socialist economic planning and Upper Silesia's importance in this mission could not be overestimated. The region, its workers and their families became the aristocrats of socialist labour. In sharp contrast to the post-war experience of the West of Scotland there was never any danger, in Upper Silesia, of demand slipping, of markets failing. The shock of developments after the 1989 collapse of communism was therefore very real and not easy, as some of our contributors point out, to accommodate. Marek Szczepański in his chapter 9 below takes up some of these themes and they are also discussed elsewhere (see Klasik, et al, 1995).

Although foreign ownership became a more prominent feature of the Scottish economy in the post war period there was no question of foreign capital involvement on any real scale in Poland after 1945 where almost all industry was in state hands. On the other hand state ownership, *the* plan and the constant drive for fast output growth generated other structural features that had a profound impact on national and regional economic development. Perhaps the most important here was the power of the heavy industry lobby and in particular the coal and steel 'lobby'. This sociological formation pitted its weight in the struggle for resources throughout the post war years and for most of that time succeeded, thus enhancing the standing of Upper Silesia among Polish regions.

Development Contrasts

The contrasts between our two regions are clear but worth summarizing across three dimensions: international, sectoral and policy aspects.

Through most of the twentieth century the international context for industry in the two regions was different. Scottish heavy industry and the economic fortunes of Scotland's West coast were firmly embedded in the UK-empire nexus in the good times to 1914 and in a de-industrializing UK in the bad from the mid 1950s. Upper Silesia and its heavy industry were, on the other hand, insulated from the world economy for nearly fifty post-war years, reconnecting only after 1989. The Polish authorities of the day were clear too in their view that reconnection should be immediate and complete. The shock was immense but reckoned to be therapeutic. There was no real thought however as to how Polish coal and steel industries would respond, sectors where product or market diversification was plainly impossible. In the end there was little alternative to the state continuing to shoulder mounting deficits in regionally concentrated and slowly contracting heavy industry.

As for the sectors themselves, the Scottish experience was one of long term decline, accelerated by the deliberate confrontation with the trades unions provoked in the early 1980s when Thatcherism reached its high point. Meanwhile, coal and steel industries in Upper Silesia could hardly produce enough under socialism and always failed to keep pace with the insatiable demands of the central planner. Then, in 1989, the Polish economic system faltered, suddenly losing energy. In 1990 economic 'shock therapy' meant an abrupt adjustment as traditional domestic and Comecon markets for coal and steel vanished overnight. In the short period 1988-90 coal extraction fell by 23 per cent and employment by 11 per cent while steel output fell by 21 per cent but its employment by only 2.6 per cent.[3] By the end of the 1990s employment in each sector was half its level of a decade earlier. In coal, the employment decline almost matched that of output. In steel it was slightly faster. The Polish experience of heavy industry restructuring in the 1990s was that it occurred at a fairly steady pace. Governments, some of which found inspiration in Thatcherite economics, opted for this 'natural' pace of restructuring, judiciously avoiding the confrontations with still strong miners' and steel workers' unions that faster adjustment would have demanded. Another major difference between the two cases is that while heavy industry has disappeared in Scotland, the enormous local coal deposits of Upper Silesia are bound to continue to be exploited for some considerable time.

Contrasts in regional policy between the two cases are also clear. In the UK, regional policy evolved from a 1960s emphasis on relocation of plants and jobs, to a 1970/80s focus on attracting inward investment. The privatization thrust of the 1980s was associated also with the promotion of 'enterprise' and the entrepreneur. It was, it seems, believed that redundant miners and steelworkers might be nurturing strong entrepreneurial instincts that the right incentives/opportunity mix would release. In the most recent period the emphasis in regional policy has switched to the need for high-level skills in high-tech local economies where, in Toothill's 1961 phrase, 'the new science-based manufactures become established on significant scale'.

Poland, on the other hand, had a simple regional policy before 1989 and practically none thereafter, until 1999. Under socialism regional policy aimed at 'equalization' and the redistribution of national product to regions. But regional policy was a very poor second cousin to sectoral policy where central ministries ruled from Warsaw. Here however, Upper Silesia, had no cause for complaint. It was home after all to the powerful coal and steel lobby to whom investments could not be denied. After 1989 the centre's powers were obviously weakened (though by no means as suddenly or sharply as many thought) but at the same time the regions had no tradition of self-activity to fall back on and through which they might have asserted influence. So-called regional policy at the national level amounted to little, its most visible aspects the 17 'special economic zones' (*specjalnych stref ekonomicznych*) where 'anti-competitive' tax reliefs to investors have dogged EU accession negotiations. Certainly, Upper Silesia, like other regions in the European post-communist economies, has attracted inward investment, especially in the motor industry and partly thanks to the special zones (described in some detail by Adam Drobniak in chapter 13 of this book). In a certain sense one can trace the shadow in contemporary Poland of regional policies used in the past in the West.

In Upper Silesia there is a strong belief that the region is different, perhaps even special, among Polish regions and enough has been said to suggest why. One rather special regional policy initiative is worth recalling. Sensing the gap that had opened up with the demise of the old highly centralized system the regional elite in Upper

---

[3] Data from *Rocznik Statystyczny*, various years.

Silesia, prodded mainly by the Solidarity trade union, argued the case for a 'Regional Contract for Katowice Province', a wide range of policy measures that would command support from Warsaw, but would be shaped locally and aimed at the economic restructuring of the region. The contract and the role of the regional elite in assembling it have been painstakingly investigated by Tatiana Majcherkiewicz. Although the contract was hardly a success in economic development terms it laid a basis in 1996, for a regional policy that seemed to have much in common with contemporary thinking on the role of partnerships in promoting regional development (discussed in the Scottish context by Mike Danson and Geoff Whittam in chapter 11 of this volume). Majcherkiewicz's persuasive account is that because the national elite in Warsaw, in the early 1990s, saw *sectors* rather than *regions* it was up to the regional elite to act to try to engage the centre in reconstruction projects. As she puts it, over 1990-97:

> the national elite concentrated on projects on the national scale, not recognizing the issues on the regional scale. This was due to the vacuum of power at the regional level ... Thus, the central elite only noticed the interests of narrow sectors of industry and was not able to put them together and see them interacting in the context of the region. (Majcherkiewicz, 2001, conclusion, p.7).

Upper Silesia was one of the few regions in Poland with a relatively strong and certainly deeply rooted sense of identity and therefore with individuals and organizations able to come to together in shaping policy. So-called 'local segments' of the regional contract brought together public and private, central and local bodies in partnerships much like those being encouraged also in Scotland today.

Lessons and Future Research?

What lessons might be drawn from a comparison of Polish-Scottish experiences in regional economic development and in restructuring heavy industry? Where might worthwhile research themes lie? Clearly the differences are many, have been touched on above and we should bear them in mind but perhaps experience permits a few observations.

First, stimulating local enterprise and awakening the entrepreneurial spirit is by no means easy. Redundant miners or steelworkers only rarely become successful entrepreneurs – either in Katowice or Motherwell. One of our seminar participants recalled the Russian workers in post-communist Sverdlovsk whose view was that 'being enterprising is fine ... until a proper job comes along'. Summing up, in 2000, the disappointing results of *Scottish Enterprise's* business birth rate strategy, one journalist observed,

> Much of Scotland's population lives in the Central Belt which used to be dominated by heavy industry and a small number of large employers – a pattern that has produced low levels of entrepreneurship in other parts of Britain (Buxton, 2000).

Nevertheless, as long as the underdevelopment of services remains great some scope for elementary entrepreneurship will exist. Sooner or later however the gap will be filled and successful entrepreneurship will need to become more sophisticated. Perhaps the remark of two specialists is worth noting in this context,

> We should give up praying for the arrival of the economic messiah, the lone entrepreneur, and build an infrastructure of networks, incubators and clusters to support entrepreneurship. (Leadbeater and Oakley, 2001).

This surely means that the contemporary push to find regional success by promoting human capital, our second general observation, is one that should command greater attention in post-communist regional transformation? The role, for example, of the 'university' as an institutional magnet for regional development seems to be relatively neglected in Poland. It is true that the private higher education sector has flourished in the post 1989 period but a shakeout is on the horizon as the demographic wave of 18-21 year olds subsides. In addition, much remains to be done in the areas of retraining and opening higher education opportunities for older age groups.

These aspects become all the more important when we understand, thirdly, that the jobs provided through inward investment often rest on a fragile base. Silicon Glen has suffered badly in the recent global downturn when the Scottish electronics industry is estimated to have lost 12,000 jobs in the eight months to May 2002.[4] How long before those assembly jobs attracted to Poland and the Czech Republic (in some cases from Scotland) move on to new locations?[5] The outward processing trade (OPT) in textiles and furniture is already shifting from Poland to the Ukraine and Russia. Even if the FDI attracted in consumer goods is likely to be robust given the size of the Polish domestic market there may come a time, as the transport infrastructure improves, when cheaper labour production sites become compellingly attractive. Inevitably, in Poland and in Upper Silesia as much as in Scotland, attention will shift to attracting better quality inward investment for which skills, education, science and technology are of the utmost importance.

It seems increasingly clear that successful, strong regional economies are likely to be vibrant in terms of social and cultural life. An active civil society probably underpins economic development and may be as badly needed, as Chik Collins reminds us in chapter six below, in parts of Scotland as in parts of Poland. The capacity of a society to generate a vision for local development and for local elites to express such visions with confidence may be a relatively unrecognized and under-researched theme in the creation of strong and competitive regions. In Scotland,

---

[4] *Financial Times*, May 28, 2002.
[5] In early 2001 the *Economist* commented, 'Scotland is suffering more than most places. Some of Silicon Glen's output is low-value assembly-line production, which Scotland imported by subsidizing foreign investment to get its electronic industry off the ground. Such work can now be done more cheaply elsewhere. Compaq is shedding Scottish jobs because it is moving computer-assembly work to the Czech Republic.' *Economist* (2001), p.32.

devolution, with the re-opening of the Edinburgh parliament, may create an opportunity to cultivate such regional development vision.[6] Upper Silesia also has a strong tradition of regional assertiveness that may find more vigorous expression in the local parliaments *(sejmik)* set up across Poland in the decentralizing reform of 1999.

It is also clear that a broadly historical approach to the analysis of economic development is important, one that helps to identify the political orientation of particular agendas and the interest structures they serve. The current failure of neo-liberal and external-investment strategies may also point to the need for a continuing development role for government.[7]

Our summary skims the surface but we hope that the comparative experiences gathered in this book may suggest other promising areas for future research and collaboration. This, and whether Scots and Silesians can learn from each other in the never-ending labour involved in building strong regional economies, is something that we leave the reader to judge.

References

Biniecki, J. Klasik, A. Kuźnik, F. (eds), (1995), *'Restrukturyzacja regionów przemysłowych. Analiza porównawcza procesów dostosowawczych w regionach Nord-Pas de Calais, Pólnocna Nadreni a-Westfalia, Górny Śląsk'* (Restructuring of industrial regions. A comparative analysis of adjustment processes in regions of Nord-Pas de Calais, Nordrhein-Westphalia, Upper Silesia), Akademia Ekonomiczna w Katowicach, Katowice.

Buxton, J. (2000), 'Business births falter north of the border', *Financial Times*, February 24.

Devine, T.M. (1999), *The Scottish Nation 1700-2000*, Penguin, Harmondsworth.

Economist, (2001), 'Silicon Glen – When the chips are down', April 21.

Evening Times, (2002), 'Fighting for a fair deal', June 26.

Foster, J. (2001), 'The Twentieth Century, 1914-1979' in *The New Penguin History of Scotland: From the Earliest Times to the Present Day*, Houston R.A. and Knox W.W.J. (eds), Allen Lane The Penguin Press, London.

Johnman, L. and Johnston, I. (2001), *Down the River*, Argyll Publishing, Glendaruel, Argyll.

Landau Z. and Tomaszewski, J, (1985), *The Polish Economy in the Twentieth Century*, Croom Helm, Beckenham.

Leadbeater, C. and Oakley, K. (2001), 'Waiting for an unlikely economic miracle' *FT Creative Business*, (Supplement), July 17.

---

[6] It is worth noting however that in the short period since the parliament opened, the Edinburgh economy has boomed while Glasgow's has fallen behind, and the problems of dealing with the effects of the loss of heavy industry are by no means solved. Glasgow's unemployment in July 2001 was 13.2 per cent, almost double the national average, *Evening Times*, (2002).

[7] Probably the most successful new area of production in Scotland in the last century was aero-engine manufacturing by Rolls Royce. When the government shifted output to Scotland in the re-armament drive before the last war care was taken to link production into existing heavy engineering supply chains. The plant was later nationalized and successfully innovated the revolutionary RB211 jet engine in the 1970s.

Majcherkiewicz, T. (2001), *An Elite in Transition: An Analysis of the Higher Administration of the Region of Upper Silesia, Poland 1990-1997*, (unpublished PhD thesis, London School of Economics and Political Science, London).

Rocznik Statystyczny, (various years), GUS, Warszawa.

Figure 1.1 Silesia 1914, the meeting point of three empires

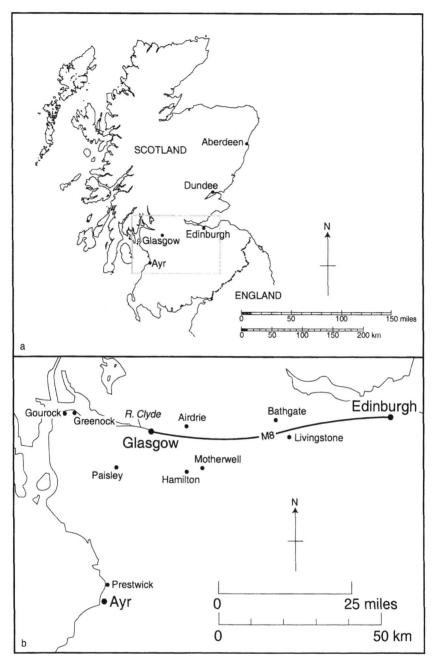

Figure 1.2 Scotland and its central belt

Chapter 2

Regional Problems, Regional Policy and Regional Well-Being – What Have We Learned in Recent Years?

Mike Danson

Introduction

This brief introduction attempts to set the arguments of the following chapters in context by providing an account of the evolution of regional problems and regional policies over the last century. It also signposts the research agenda for the volume as a whole. Although fairly discursive, and by no means comprehensive, the intention is to inform the exploration and debate on the restructuring of heavy industry regions by drawing on the literature and the general developments of the last few decades. An extensive catalogue of such research exists (see for example Armstrong and Taylor; 2000; McCann, 2001; Danson and Whittam, 1999) but in some cases it tends to neglect the need for underlying theoretical frameworks for analysis and in others findings tend to have withered with time and experience[1].

We organise this discussion beginning with a review, in the next section, of the patterns, causes and environment of regional problems in the UK. We explore in particular the congruence of the problem regions in the 1920s and in the early part of the new millennium. This points to deep-rooted, past dependent causes of regional disparities. Recent discussions of regional problems and policy have focused and drawn on theories of 'endogenous growth'. We review those developments and the interests involved in the institutions created to foster regional development. Such institutions typically are co-ordinated through formal partnerships and this leads us naturally to consideration of the significance of governance and government in determining how policies are pursued within regions. Finally, some conclusions and areas for further discussion are proposed.

---

[1] See for example the optimistic initial conclusions on the success of measures addressing the catastrophic decline of the UK coalfields (*Manual – Social Crisis Management in the Coal and Steel Industries. European Models and Experiences,* Hans-Werner Franz, Commission of the European Union, 1994) as compared with more realistic recent evaluations.

Regional Change and Regional Peripheries

It is striking that in the UK, as in many other developed economies, the same regions which suffered massive unemployment and industrial dislocations in the depressions and recessions of the period after the first world war are those which faced consistent problems thereafter. In such regions living standards and industrial growth rates have lagged behind the national average for almost ninety years, suggesting that systematic underlying forces are at work. Indeed, it has been argued that the traditional industries of the north and west of Britain were dependent on the Empire and then Commonwealth for secure markets for the capital goods they specialised in. Loss of Empire inevitably led to disruption and unemployment in those regions. The 1920s and 1930s witnessed the first cycles of 'cumulative causation'. Although the second world war and the long boom of the post-war period obscured some dimensions of decline, the coal, steel, shipbuilding and heavy engineering industries of Clydeside (Scotland), South Wales, and Tyneside in the north east of England continued to lose competitiveness. While those areas had led the world into the industrial revolution they were in almost perpetual crisis throughout the twentieth century. In the UK, the nationalisation of most of those sectors in the late 1940s and again in later periods of Labour government rule illustrate not so much their strategic economic role but rather the fact that they required protection after persistent underinvestment by their private sector owners.

The capital goods sector was especially prone to cyclical changes in demand. The dominant employers in Clydeside, South Wales and Tyneside were concentrated in precisely those industries: any sectoral downturn inevitably triggered severe regional problems. Indeed they came to be accepted almost as a natural phenomenon. Later, over 1945-80, as governments changed, privatisation followed nationalisation (sometimes with a bout of re-nationalisation) but the backdrop was one of steady, overall decline in the heavy industry sector. The era of Thatcherism and monetarism accelerated that process. Very rapid and widespread closures of coalmines, shipyards, engineering and steel plants in the 1980s led to massive industrial restructuring at a pace and depth not experienced before in peace time. The dominance of many of those sectors locally, amplified through clusters of related and supplier industries around them, meant that the problem regions saw their prospects worsen both relatively and absolutely. It is forecast that by 2010 as few as 11 per cent of workers in Scotland will be in manufacturing. In 1921, by comparison, one in six was employed in mining alone. Scotland is typical of such old industrial areas where the past shapes immediate problems and a clear path dependency affects the course of regeneration.

Decline was exacerbated by a number of other elements. In particular, the loss of formerly secure Empire markets was paralleled by the expansion of Europe's significance for British trade. The traditional industrial regions, on the wrong side of the country, on the periphery of the continent and tied into redundant sectoral links, were not helped by the UK's entry into the European Community. Though the arguments are complex here, the interaction of a number factors needs to be taken into account: the growing dominance of competition policy over all other EU policies with the subordination of regional and industrial policies especially important; the related loss of many traditional regional and industrial policy instruments, including devaluation, nationalisation, national public procurement practices which formerly would have been marshalled to address the developing crisis in the regions; the ongoing centralisation and concentration of political and economic power in corporations and higher levels of government located in capital and 'world' cities. Each factor contributed to the progressive and cumulative diminution of regional power to overcome loss of competitiveness The more recent activities of the World Trade Organisation (WTO) have extended and focused the deleterious effects of these large scale changes on old industrial regions, accelerating decline while removing traditional regional and national mechanisms to alleviate unemployment and loss of output and to promote regeneration.

Shifts in Regional Policy

Thus, local levers of control were being lost through the closure of regionally based industries, at the same time as the Keynesian demand management system was being dismantled, its essential spatial elements downgraded. Unemployment created by industrial change led to concentrations of poverty and deprivation in the north of the UK, increasing the costs of managing decline and undermining strategies for improvement. New economic activity was locating in the south east, around the capital and closer to mainland Europe, in new towns and other growth poles.

Through the periods when governments had attempted to encourage private capital as well as the nationalised industries to locate in the regions, there was a strong reliance on foreign direct investment (FDI). But important questions were being overlooked. Were the benefits of FDI sustainable in terms of jobs and incomes? Were linkages being established between incoming plants and local suppliers? Was the regional economy becoming diversified or was a cluster of dominant sectors being replaced by a new non-indigenous set of branch plants? To many it looked like the regional economy was not so much being *restructured* as being *underdeveloped*. In any event, as transnational capital became even more footloose and mobile in the 1980s and as new locations joined the competition for the plants which were available, so the relative attractiveness of the periphery of the UK was undermined. As restrictions on applying traditional policy instruments to restructure old industrial areas evolved, local and regional partners became more closely involved in employment and economic policies in their areas. This enforced reorientation of local economic development away from top-down relocation of industry and attraction of inward investment towards the promotion of local strategies and solutions was not inconsistent with the national, if not global, moves to promote small and medium-sized enterprises, entrepreneurship, new firms and other forms of indigenous development.

Endogenous Growth Theory

Gradually over the last quarter century traditional demand management approaches at the regional level have either failed, appeared to fail or been rejected. They have been replaced by so-called endogenous growth programmes and policies, including a focus on the supply side of the economy and the labour market in particular. With a parallel move away from investment in physical infrastructure and towards expenditure on training, labour mobility and other employment schemes, the specific needs of particular localities were downgraded in favour of national strategies for skills and competencies.

Based on notions of the industrial district, clusters and networks, founded on ideas of trust and cooperation, learning and the knowledge economy, a new form of intervention emerged. Entrepreneurship was to be encouraged through support for new small businesses, reduced bureaucratic controls over enterprise, and more expenditure on training and human capital development, with the gamut of government policies and programmes more business-led than in the past.

With strategies to address the Business Birth Rate, clusters and the knowledge economy, Scotland was at the forefront of many such initiatives. Indeed, it was often the conduit for the latest ideas from America in working with the market to restructure and regenerate the regional and local economy. As will be seen later in this volume many of those policies have been less than successful in their Scottish implementation, suggesting that history, industrial and social legacies are not irrelevant in determining how transferable regional solutions can be.

Earlier support for manufacturing at the regional level gave way to encouragement to service sector development, the latter often having been actively discouraged in the past in restructuring strategies. The emphasis on technology and technical progress, which underpins much of the endogenous growth theory literature, also tended to be applied across the country or continent, regardless of initial regional disparities. With most innovations being introduced in the core of the economy first, technological progress led to increasing regional differentiation. Jobs in traditional industries in the problem regions were destroyed with new employment moving closer to the capital city and associated centres of power. The other main planks of the new economy - training and education - were again not spatially indiscriminating. Deskilled workforces and redundant industrial landscapes were not conducive to new inward investment, to new competencies being nurtured nor to new local entrepreneurship. Cumulative causation cycles and past dependencies were significant factors in curtailing the applicability of many models and approaches to the regeneration of the old industrial regions but all too frequently they were neither recognised nor acknowledged. This led increasingly to communities themselves being blamed for mass unemployment, low rates of new firm formation and lack of retraining.

Institutions, Development Agencies, Regional Governance and Government

This translation of national regional and industrial policies into actions for local economic development came to involve a range of actors and social partners.

Although the focus for implementation of projects and programmes was to become the locality - and often this meant the community or neighbourhood - a range of governmental and other agencies was usually involved in a partnership approach to delivering economic development and training services. This was especially the case where deindustrialisation and associated regional problems were deeply rooted. In Scotland, Wales and Northern Ireland regional development agencies (RDAs) were created to target UK interventions specifically on those areas, while local authorities had evolved an understanding and accommodation with RDAs to make actions more effective and efficient. The perceived advantages of using an organisation at arms-length from government itself are widely accepted (Halkier, Danson, and Damborg, 1998) and most countries utilise this approach to delivering policy initiatives. While lack of institutional capacity or thickness may be at times a difficulty, a more recent problem is a perception of overlap and duplication in the crowded economic development landscape. How this myriad of players and agencies is coordinated and managed has become a new area for concern.

A peculiar local example of the evolution of partnership involved the formation of an independent executive to co-ordinate and manage operations and programmes. Ultimately, as regional restructuring efforts engaged EU structural funds, this 'Strathclyde European partnership model' was increasingly adopted across the continent. An environment of consensus and corporatism in Scotland promoted such an approach to regional economic development, one that seemed eminently compatible with the strategic thinking of the European Commission. However, many partners in this arrangement are institutions at arms-length from their appropriate layer of government raising questions over democratic accountability and responsibility. This suggests that the involvement of directly elected representatives is essential to ensure legitimacy of strategies and programmes. Both theory and experience suggest that there is an appropriate role for the different levels or jurisdictions of government in delivering local and regional economic development, with local, regional national and European all able to claim that some aspect of regional policy is most successfully delivered if they are concerned directly in the policy process (Armstrong, 1995).

Uneven Development and Cumulative Causation

Despite 80 years of interventions, innovations and partnerships, it is clear from the vast literature on economic development that regional disparities persist. The market plays its part, causing regions and nations to develop at different rates, with some always lagging while others prosper. Such effects can become self-perpetuating, especially in a new world of globalised markets, regulations and corporations. With fewer instruments to manage the national or regional economy, innovative and dynamic activity is increasingly attracted to the core. The market, therefore, fails most dramatically at the periphery, with the poorest regions and communities unable to turn their economies around quickly or in sustainable ways. As the elements of cumulative causation become embedded in the regional economy power also moves progressively to higher levels.

An important parallel is that, as Scotland has suffered at the western periphery of the continent, so the eastern margins of the enlarged EU may also be faced with a future which promises restructuring and redundancy, but with even fewer policy instruments to address long term structural difficulties. The evidence seems to suggest that when European output increases overall, then EU regions may experience convergence in GDP, employment and wealth. If EU enlargement means faster rates of growth then sufficient surplus to promote activity in the new periphery may exist. If not, then the gap with the centre of Europe will at best take many years to close and may indeed widen.

The recent history of Ireland apart, even where accession countries grew fast as in the 1980s and 1990s divergence within nations increased. Generally, peripheral and older industrial areas have been most affected by the challenging national agenda posed by membership of the EU club: the need to adapt to a customs union and the single European market. The agenda is made even more challenging by the emergence of *Euroland* and the desire of most countries to be in it. The restructuring implications are made more severe with the loss of economic and monetary independence. Former coal and steel areas have been hit particularly hard with few on the periphery achieving successful models of sustainable and rapid regeneration. They are the regions least able to take advantage of regional and national 'endogenous growth policies' with their focus on enterprise, new technologies and a dynamic and flexible economy. In those regions the legacies of the past remain very real.

Conclusion

A response to the problems of path dependent relative decline has been pressure across Europe to (re-)introduce more local political control over the economy and society. This has been strongest outwith the federal states, partly because of a well founded belief that countries such as Germany and small independent nations have been better equipped to meet the challenges of the latter part of the last century than unified monolithic states. The stateless nations (for example Catalonia, the Basque country, Scotland, Wales) within over-centralised EU countries have led the way in demanding greater autonomy as a way of emulating the enhanced flexibility and control enjoyed by the smaller member states and the *länder*. That several of the communities seeking change are dominated by some of the oldest industrial areas in their respective countries, if not the world, raises important issues for the next wave accession states also.

If the lesson of successful industrial districts and clusters is to demonstrate the significance of trust and co-operation, as well as a coherent and inclusive society, to economic regeneration and effective restructuring, then the those Central and East European countries soon to join the EU may need to address wider regional development issues than first envisaged. The market alone cannot ensure the conditions for sustainable development in areas facing massive industrial change. Institutional and policy innovations will be essential elements in this process. Yet the WTO, EU with their related regulations and agreements restrict the scope for policy development. The accession states, and their weaker regions, may need much more than the existing international policy-making architecture permits if they are to reap the anticipated benefits of EU membership.

This volume tries to set the scene, in a Polish-Scottish context, by encouraging the reader to compare regions, levels of development and policy evolution across two rather different situations, where industrial restructuring problems have been separated in time, yet which share some common difficulties of peripheral, in EU terms, location.

References

Armstrong, H. (1997), 'Regional-level jurisdictions and economic regeneration initiatives', in M. Danson, G. Lloyd, and S. Hill (eds), *Regional Governance and Economic Development*. London, Pion.

Armstrong, H. and Taylor, J, (2000), *Regional Economics and Policy,* 3rd edition, Blackwell, Oxford.

Halkier, H, Danson, M. and Damborg, C. (eds), (1998) *Regional Development Agencies in Europe*, Jessica Kingsley Publishers, London.

Danson, M. and Whittam, G. (1999), 'Regional Governance, Institutions and Development', in *The Web Book of Regional Science*, Regional Research Institute, West Virginia University, http://www.rri.wvu.edu/WebBook/Danson/contents.htm

Franz, H. W, (1994), *Manual – Social Crisis Management in the Coal and Steel Industries. European Models and Experiences,* Commission of the European Union, Brussels.

McCann, P. (2001), *Urban and Regional Economics*, OUP, Oxford.

PART II

ECONOMIC TRANSFORMATION AND RESTRUCTURING – FROM HEAVY INDUSTRY TO WEIGHTLESS ECONOMY?

Chapter 3

Can Regional Policy Meet the Challenge of Regional Problems in Poland?

George Blazyca, Krystian Heffner and Ewa Helińska-Hughes[1]

Introduction

Many of the contributors to this volume draw attention to the decentralizing reform of state administration that was hotly debated in Poland over 1997-98 then introduced from 1 January 1999. That reform was undoubtedly the single most important measure with a bearing on regional policy of the first decade of post-communist transformation. The reform had several aims. It was a much-delayed response to the decentralizing expectations of 1989-90 in the political and social sphere. It was also designed with an external dimension very much in mind and associated with expectations of a different sort, that structures had to be created to weld Polish regions into potential EU funding streams. Finally, it was an opportunity to give a genuine fillip to domestic regional policy developments. Our aim in this chapter is to comment on the success of that reform in its first period of implementation. We surveyed the Polish professional literature, government and authoritative media in the period through to late 2001 and early 2002 to try to tease out a sense of the major issues in contemporary Polish regional policy as well as perceptions of problems for the period ahead. While the 1999 decentralization did establish a new institutional framework our impression from our survey is that the regional policy that would utilize and animate such a framework remained underdeveloped and faced potentially very serious problems in the light of EU accession. The greatest danger is that emerging regional imbalances of the later 1990s may be exacerbated by EU entry, as relatively successful regions connect well with Brussels and Warsaw but others do much less well and continue to slip. Policy, in a sense, lags institutions. New administrative structures exist but the policy to activate them is, at least in part, missing. In short, some Polish regions stand to do very well post EU accession but others may do badly with divisions across the country becoming even deeper.

---

[1] This chapter draws on and follows closely a previously published paper which appeared in *European Urban & Regional Studies*, (9) 3, July 2002. The authors are grateful to Sage Publications Ltd for permission to use that material here.

Fault Lines in Polish Economic Space

For much of the post 1945 period notions of 'balanced', 'harmonious' and 'equal' development meant that the fault lines running through Polish economic space were either hard to see or not much discussed (OECD, 1992; Gorzelak, 1998). The passage of time however brought them into stronger relief. Three are especially striking and have continued to exercise a strong influence in the post-communist period. One was sectoral where industry, especially heavy industry, metal production, engineering and shipbuilding were accorded special status in socialist modernization. The regions where such activity was located benefited accordingly, none more so than the heavy industry conurbation of Upper Silesia.[2] The second was the urban/rural divide with associated ownership and cultural dimensions. The third was the particularly steep geographic gradient between eastern and western regions, between 'Poland B' and 'Poland A' as it has been rather indelicately put. We consider each in turn.

Troubled Sectors, Troubled Regions

It was widely expected in the early 1990s that serious transformation problems would quickly appear in the industrial conurbations dominated by overblown and uncompetitive heavy industry. Regions with the 'wrong' economic structure could expect problems. Upper Silesia, commanded much early attention because it contained almost all of Poland's basic heavy industry. As post-communist economic transformation started in 1989 the region had 64 of Poland's 66 deep coalmines and one of the two major Polish steelworks. In 1989 it accounted for 96 per cent of Polish coal production and over 50 per cent of national steel output (*Rocznik Statystyczny Województw*, 1997). Large-scale labour shedding and inward investor hesitance, not least because of environmental issues, were greatly feared in the early 1990s. While other regions (Łódź with its concentration of textiles is another clear example) faced similar problems Upper Silesia looked to have the worst of it because of the sheer concentration there of 'sunset' sectors. Facing a sudden collapse in traditional markets for its coal and steel, and with immense environmental problems, Upper Silesia was widely thought to be the obvious potential flashpoint (OECD, 1992; Gorzelak, 1996, 1998). Some observers feared that a sectoral implosion in Upper Silesia would provoke first a regional and then a wider social and political crisis as the

---

[2] The concept of 'region' has many interpretations from a functional 'geographic space with administrative capacity for regional development' (Kafkalis and Thoidou, 2000) to a 'socially and culturally constructed category ... emerging from communication and social action' (Jensen and Leijon 2000). In addition, it has become possible, indeed fashionable, to identify 'intelligent regions', those that are purposeful, able to exercise some control over structural transformation (Kafkalis and Thoidou, 2000). In the discussion that follows 'region' is understood principally in administrative terms. Poland was, from 1973 to 1998 a country of 49 regions (voivodship – *województwo*) then, from 1999, 16 new regions replaced earlier structures. For the purpose of this paper it is one or other of those two sets of regions that provide the underlying data permitting some generalization on developments in Polish economic space.

former aristocrats of the working class – the miners and the steelworkers – were forced to submit to new economic realities. In the event, even the most economically liberal of Polish finance ministers could make only small headway in the 1990s in imposing hard budget disciplines on mines and steelworks. Such restructuring of Silesian heavy industry as did take place was the result, for most of the period, of 'natural wastage' rather than explicit programme design. It was only from 1998, with World Bank and other external financial support, that miners were offered (and accepted) voluntary redundancy on a large scale. Between May 1998 and May 2001 Polish coal mining employment fell from 243,000 to 153,000. Of the 90,000 jobs lost, 28,000 were due to natural wastage and 62,000 involved early retirement and voluntary severance deals (*Gazeta Wyborcza* 2001a). In the steel sector over-capacity has created great awkwardness in the run up to EU membership where the latest (summer 2001) agreement between Warsaw and Brussels allows the Polish government to subsidize the industry in the context of reducing its employment from 40,000 to 30,000 by 2003 and holding output steady at 10m t per annum (Bielecki, (2001a, 2001b).

Restructuring heavy industrial regions is of course something that many western countries have experience of but surprisingly little attention seems to have been devoted to identifying any lessons for post-communist transformation. From a developmental perspective one striking, although perhaps not unexpected, feature of Upper Silesia is weakness in generating entrepreneurial activity, its equivalent of the poor business birth rate much discussed in Scotland (*Scottish Enterprise*, 1993, 2001). A well-informed journalist reviewing the failure of Scottish Enterprise's business birth rate strategy noted that, 'much of Scotland's population lives in the Central Belt which used to be dominated by heavy industry and a small number of large employers – a pattern that has produced low levels of entrepreneurship in other parts of Britain' (Buxton, 2000). Polish researchers have observed a similar phenomenon. Surażska notes for instance that the heavily industrialized Silesian urban centres, despite populations with higher than average family incomes, display little enterprise, 'a fundamental problem closely associated with mining restructuring' (Surażska, 1999). The Silesian regional problem has clearly not disappeared and badly needs an active transformation policy.

If Upper Silesia is rightly regarded as a special regional development case because of its exceptional concentration of heavy industry there were many other more 'routine' examples where monolithic industrial development created problems from the post-1989 decline of the Łódź textile industry to the difficulties faced in the 'one-plant towns'.

Urban-Rural Tensions

The urban-rural dimension continues to have an impact on patterns of regional development. Rural areas dependent on agriculture suffer as much from underdevelopment as from unemployment (OECD, 1996). However, most large urban centres, especially in the west, eventually did recover following an initial early 1990s transformation shock. When foreign direct investment started to flow its natural magnets were those western urban centres (Gorzelak, 1998). Investors were attracted by markets, by reasonably well-developed infrastructure, by good prospects for privatization of better quality assets and by ample supplies of cheap labour. On the other hand predominantly rural regions were in trouble, handicapped by weak infrastructure, lacking the important metropolis effect that comes from association with large centres of population, with poor privatization prospects or dominated by a highly fragmented agriculture that was about to lose traditional subsidies. The collapse of state farming in the north-west, for example, was about to push that region into severe difficulty.

In the 1990s the urban-rural development gap widened. Relative price movements (the 'terms of trade' between town and country) give one indication of those developments. In 1994-95 agricultural produce prices grew slightly faster than prices of industrial inputs purchased by farmers but in the years that followed the 'prices scissors' opened to the great disadvantage of the former, the relative price index declining from 101.8 in 1995 to only 91.5 in 1999 (*Rocznik Statystyczny 2000*). Farmers' incomes took the brunt of this adjustment. Farming households in 1999 had an average income per person equal to only 73 per cent of the national average as compared to 94 per cent in 1995 (*Rocznik Statystyczny, 1996, 2000*) and, not surprisingly, the late 1990s saw a number of nation-wide and well-organized farmer protests. These reached a crescendo in 1998 and in January 1999 when the farming union (soon to be a political party) *Samoobrona* led by Andrzej Lepper became particularly prominent.

Polish agriculture, with 18.8 per cent of national employment but only 3.3 per cent of gross valued added, also remains a highly sensitive and critical matter in Poland's bid for EU membership (Commission of the European Communities, 2001, p.51). Writing shortly after the December 2000 Nice summit Laza Kekic of the London based Economist Intelligence Unit observed that Poland 'is the most difficult to absorb – because it is large, poor and has a relatively backward agricultural sector' (Kekic, 2000). The countryside, more generally, suffers from a range of serious problems: hidden unemployment, poorly developed infrastructure as well as limited educational and training possibilities. Rural areas are also highly differentiated in terms of preparedness for EU accession with regions in the south-west and north in a much better position to do well out of EU membership than those in the centre and north-east. Their advantages stem from having better general economic infrastructure with denser urban communications and cultural connections, a better qualified labour force, greater labour mobility, richer educational possibilities, a greater share of non-agricultural employment even in rural areas and a better farm structure with larger units (Hunek, 2001; Rosner, 2001). In general however the preparedness of Polish agriculture for EU accession looks weak. The European Commission's annual assessment of Poland's readiness for accession concluded, in November 2001, that as regards agriculture a 'coherent strategy for the sector is still lacking' (Commission of the European Communities, 2001). It also noted that no progress had been made in the previous twelve months in such matters as classifying less favoured areas for rural development support within the SAPARD (Special Accession Programme for Agriculture and Rural Development) scheme.

The East/West Divide Becomes More Complex

The east-west divide has become more 'fuzzy' during the 1990s but it remains visible in almost all socio-economic indicators from income/capita, unemployment rates, FDI inflow, conditions of economic and cultural infrastructure, entrepreneurship, system transformation as well as wider measures of 'social' or 'civic' dynamism. In general regional differentiation appears to have increased along the east-west axis during the 1990s although the picture also became more complex. The collapse of state farming brought, for example, greater deprivation to rural northwestern Poland while the boom in informal trade created new business opportunities in some frontier regions in the east. Some cities and towns close to the German frontier also received a substantial 1990s boost from cross-border trade and co-operation (Stryjakiewicz, 1996; Heffner, 1998). Areas close to Szczecin in north-west Poland, Zielona Góra in the central-west, Opole and Wrocław in the south-west and even Katowice, traditionally linked to the German labour market through migration, also benefited (Heffner and Solga, 1999).

The 'blurring' of the east-west divide is partly due to choice of development indicator, time period, level of aggregation, the decision on whether to include or exclude the Warsaw region and the nature of summary statistics employed (absolute or relative range, coefficient of variability and so on). Some authors detect clear signs of a sharper differentiation in the post-communist 1990s. Mierosławski and Jakubowski (*Rzeczpospolita* 1998), for example, show that in 1992 the Płock voivodship (to the west of Warsaw) had a level of GDP per capita some 2.5 times greater than Żamojski (in the south-east on the Ukrainian frontier) but by 1995 that ratio had reached 3.5:1. Moreover, according to the same authors only nine of Poland's 49 regions (voivodships) had, in 1992, GDP per capita less than 75 per cent of the national average but by 1995 that number had increased to 18. If the data are recast into the mould of the (16) larger regions that from 1999 replaced the smaller voivodships differentiation is, expectedly, less severe but still on the increase. Orłowski reports the GDP/capita spread widening from a maximal 1.9:1 in 1995 to 2.2:1 in 1997 between the Mazowsze region with Warsaw at its heart and Świętokrzyski just to the south (Orłowski, 2001). In a thorough review of the evidence on regional differentiation one expert concludes that there is no unanimity of opinion among Polish researchers, that judgements are sensitive to choice of measure used, and that moving from reliance on GDP/capita would bring a wider understanding of the changes taking place in Polish economic space (Dziemianowicz, 1999). There seems to be little doubt however that large urban centres (and they are mainly in the west or centre of the country) – Warsaw, Poznań, Wrocław, Kraków, Gdańsk, Łódź and the former socialist industrial dynamo in Katowice – all performed relatively well through the 1990s. But for cities in the 'eastern arch' from Olsztyn, to Białystok, Lublin and Rzeszów 'post-communist transformation' was a much harder experience (Rykiel, 1995). With a few exceptions the east still remains a peripheral region in terms of economic development even if the mosaic is, in the new millennium, more intricate. Appendix Table 3.2 gathers recent GDP and unemployment data by region.

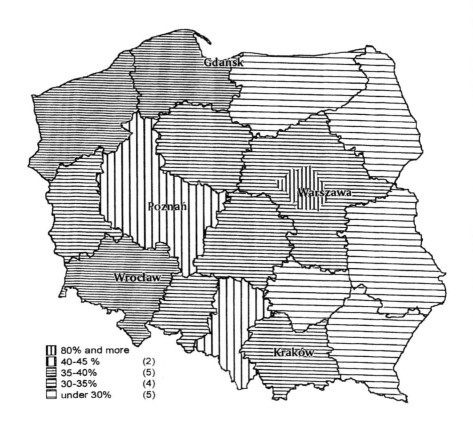

Source: From Orłowski (2001)

Figure 3.1   GDP Per Capita by currency purchasing power in 1997
(Average Level of EU 15 = 100)

From the late 1990s independent investigators have looked more closely at regional patterns of development and all confirm a continuing east-west divide. The Gdańsk Institute for Market Economics (*IBnGR*) was one of the first of the independent research centres to probe Poland's economic space systematically and its now regular surveys of the economic attractiveness of cities, medium- and small-sized towns give ample evidence of the continuing durability of that spatial division. Its 1999 survey of 260 district (*powiat*) capitals yields typical results (Swianiewicz and Dzemianowicz, 1999). As in all such *IBnGR* surveys towns are placed on a seven point (A to G) scale on the basis of ranking across a wide range of socio-economic indicators. For reasons of delicacy little detail is given (in press reports) on the lower end of the scale but enough is said to reinforce the general development impression outlined above. As the *IBnGR* puts it:

> much research confirms that the least developed regions are in the east. Half of class A towns are close to the western frontier while more than half of the least attractive are close to the eastern border ... no class G town can be found close to the western frontier and no class A town exists on the 'eastern wall' (*ściana wschodnia*).

Size matters too. Successful towns are larger, the least successful small. Again, the figures are striking: among large towns (with over 50,000 people) some 60 per cent are A/B types and only 3 per cent F/G but for small towns (with fewer than 15,000 people) only 3 per cent managed to get into the A/B category while 50 per cent were F/G.

	core and development poles
	development axis
	semi-peripheral regions
	peripheral and declining regions
	lagging regions

Source: From Rykiel (1995).

Figure 3.2 Core, Peripheries and Lagging Regions at the End of the 1990s

Eastern regions of Poland are also distinguished by particular demographic and employment structures, their much greater shares of pre- and post-productive age groups generate correspondingly greater demand for social services. They are mainly agricultural. Indeed in the three voivodships (Podlaskie, Podkarpackie and Lubelskie) along the eastern wall agriculture accounted in 1999 for 45 to 51 per cent of total employment (*Rocznik Statystyczny Województw*, 2000).

Economic success in Poland in the 1990s thus depended mainly on being in a large town, close to large centres of population in the west of the country. By the time the Gdańsk Institute reported again in 2001 there was, as to be expected, relatively little change (*Gazeta Wyborcza*, 2001a). Several eastern regions still had no class A/B towns but one (Podkarpackie) had picked up, winning an 'A' for 8 towns, an achievement attributed to education and labour quality improvement as well as the nourishing of a 'civil society encouraging inward investment'. It was becoming clear that features other than simply location were at work in shaping development. More general transformation capabilities were also playing an important part. One author (Surażska, 1999) writing in late 1999 noted that her research on development potential across *powiat* towns showed a strong correlation between market economy and civil society. Entrepreneurship in the business sense was often associated with vigorous development of non-profit organizations and lively and committed 'self-government'. Releasing the civic energy associated with the latter was of course to be a key feature of the 1999 reform.

Polish Regional Policy to 1999

In communist Poland regional policy was targeted towards 'equalization', its principal instrument investment: if the centre believed it had to do something for a weaker region it was enough to take a new plant to the town concerned. The planner was in complete charge and it seemed that there was little that could not be achieved simply by decree from Warsaw. In practice however regional policy achievements were less impressive as the persistence of the 'Poland A/Poland B' syndrome testified. Central planning and the logic that drove it created single-plant towns sometimes, as in the case of the Lenin steelworks in Nowa Huta by Kraków, with social engineering as much an objective as its mechanical counterpart (Stenning, 2000). National needs dominated all as the ruthless exploitation of natural resources clearly showed. As time passed industrial development and infrastructure imbalances became increasingly evident. In communist Poland no motorways were built, waterways were neglected (a legacy that must have exacerbated the flooding problems in 1997 and 2001), the telecommunications system was grossly underdeveloped and the network of higher educational establishments was skewed in favour of the needs of industry and the national plan. Perhaps however the most important weakness of all, with long-lasting effect, stemmed from centralization: regions run from Warsaw had no institutional capacity for self-managed development. Lack of such capacity is by no means a novel phenomenon within the EU where in the course of previous enlargements peripheral areas, for example in Greece, have been generously supported to establish appropriate institutions to connect better with EU funded infrastructure investment (Kafkalas and Thoidou, 2000).

In the period immediately following the collapse of communism in 1989 macroeconomic stabilization was the priority and it is not surprising that the 'Balcerowicz Plan', hardly mentions regional policy and concentrates mainly on structural changes (Gorzelak, Jałowiecki, Herbst and Roszkowski, 1999). In so far as it had any regional dimension it paid more attention to arrangements, in Poland's contemporary two tier regional structure, between the municipality (*gmina*) level and the centre, rather than to the voivodships (Hausner et al, 1995; OECD 1992). Economic space in post-communist Poland was being spontaneously reshaped by the market and regions were left to benign neglect. According to one well-known Polish regional specialist sectoral policies dominated regional policies in the early 1990s (Gorzelak, 1996). Arguably a stronger formulation is justified: it might be suggested that centralization suddenly collapsed into 'voluntarism' with any carefully considered decentralization some way off. The only governmental department conducting any explicit form of regional policy in the early 1990s was the ministry of labour and social policy in its attempts to slow unemployment growth. Later, the mid 1990s saw some declarations but few practical innovations in shaping or promoting regional policy. The centre-left parties that took power from Solidarity in 1994 were certainly not afraid of 'activism' and by no means doctrinally opposed to interventionism but nor were they committed decentralizers. The only significant regional policy innovation of the mid 1990s was the controversial experiment in setting up 'special economic zones' (*specjalne strefy ekonomiczne – SSEs*) offering tax breaks to investors in regions with particularly severe problems, an experiment that was to cause immense difficulties in EU accession negotiations, and with ambiguous economic impact (Kryńska, 2000). The mid 1990s SLD-PSL coalition government established 17 such zones, of which 15 became active, extending tax holidays to the year 2017 to some 700 investors and creating 28,609 jobs to the end of the year 2000 (Oktaba, 2001). The European Commission argued that the zones breached competition policy rules but by late 2001 it seemed that a compromize was close, given the genuinely weak condition of the areas concerned, which would allow the Polish authorities to incorporate the SSEs in a broader framework of regional policy (Bielecki, 2001b).

Regional policy gained greater prominence when the Government Centre for Strategic Studies (*Rządowe Centrum Studiów Strategicznych - RCSS*) was set up in 1997 and given responsibility for, among others, conceptual developments in spatial policy. But the RCSS, as a 'think tank', had no operational duties and little influence over the regional activities of government departments. Much more significant was the EU's decision in 1998 to open membership negotiations with Poland. This triggered a general change in the policy-making climate in Warsaw (where in 1997 another Solidarity coalition government had been installed), one that would be dominated by the need to conform to the *acquis communautaire*. Regional policy immediately began to command much greater attention. A department for regional development was set up in the economics ministry. But by far the most important development was the administrative decentralization promised by the new government as part of the raft of reforms (the others were in pensions, health and education) it claimed had been damagingly postponed by its predecessor. The 1999 regional reform, after considerable and heated political debate, replaced the existing 49 voivodships with 16 new 'super regions'. It also reinstated a middle tier of local government in the shape of the district (*powiat*) and continued to rely on the municipality (*gmina*) for delivery of services at the lowest level of administration. By the end of the 1990s, amidst all of these organizational-administrative innovations, an old dilemma of regional policy, but one that could be easily overlooked in the communist period – efficiency versus equality – was clearly also back on the agenda.

The 1999 Reform – Design, Experience, Fears

Design

Two crucial elements lay behind the 1999 reform: it delivered a decentralizing promise made at the time of the collapse of communism ten years earlier in 1989 while equipping the country with administrative structures that would facilitate EU entry and access to the EU's structural funds. The birth of new regions and the 'empowerment' of the middle-tier districts would, it was hoped, release independent regional and local energy that would promote economic and social development. Exactly how these aims would be achieved was not clear at the reform launch. It was plain however that administrative decentralization could not end only with the publication of colourful new maps or by ceremonially naming 16 (or as it turned out, 18) regional capitals. Decentralization had to imply the transfer of resources as well as responsibilities. At the same time, another aim of the decentralization process was to break up the sectoral/branch state structures inherited from the command economy (Gilowska, 2000).

In May 2000, some 16 months after the reform was launched, the government attempted to fill in the missing details in a law on regional development (Dziennik Ustaw, 2000a). This usefully set out broad development aims and proposed new institutions and instruments to achieve them. According to the legislation, regional development has three goals. It should 'improve the quality of life'; promote regional competitiveness (*konkurencyjność*) and 'equalize differences in development levels as well as citizens' opportunities no matter where they live'. Further, more specific regional development tasks were outlined:

1. Promotion of enterprise, the SME sector, innovation and technology transfer.
2. Restructuring public services and local economies in the spirit of balanced (*zrównoważony*) development.
3. Creation of permanent new jobs.
4. Investment in infrastructure, especially transport.
5. Development of education, especially for adults.
6. Promotion of regional culture.
7. Environmental enhancement.
8. Institution building to support local development (*pobudzanie aktywności*).
9. Research to underpin regional development.

Meanwhile an integrated planning framework for regional development was elaborated introducing one important new instrument, the 'regional contract' (*kontrakt wojewódzki*), between centre and voivodship, as a critical mechanism in achieving the objectives outlined above. EU membership issues, especially linkages to EU funding streams, had a clear impact on domestic institutional innovation. A medium term National Development Plan (*Narodowy Plan Rozwoju - NPR*), roughly coinciding with the six-year EU budget horizon, would be generated from a National Strategy for Regional Development (*Narodowa Strategia Rozwoju Regionalnego - NSRR*) as well as development plans locally produced by the 16 regions. The first NSRR, covering 2001-06, was approved in late 2000. At central government level the economics ministry (*Ministerstwo Gospodarki*) had, over the mid to late 1990s, emerged as the *de facto* co-ordinator of regional policy.

So much for reform design. What of the early experience? What does it tell us about the development of regional problems and regional policy in Poland? A lively discussion in the country helps to throw light on some of these issues.

Experience

As we noted earlier releasing civic energy in lively and committed self-government was to be a key feature of the 1999 reform. What was achieved?

Some observers have expressed doubts as to the capacity of the new structures genuinely to support 'bottom up' conceptions of local and regional development. The creation of 16 large regions *appears* to satisfy the need for reasonably strong regional development agents permitting both real decentralization and smooth linkage to EU structures. However the first experience in setting up 'regional contracts' suggests, say critics, that the government hands over resources to support mainly the local delivery of *national* programmes (for example in health and education) where regionalism barely figures (*Gazeta Wyborcza*, 2001b). This *is* a decentralization albeit highly managed and tightly controlled. Some 80 per cent of the finance available for regional contracts was to be allocated on a population basis, with 10 per cent for regions where GDP per capita is below 80 per cent of the national average, and another 10 per cent for regions where unemployment exceeded the national average by 50 per cent for at least three years (*Dziennik Ustaw*, 2000b). As might be expected, for some, this represents too paltry a start towards regional equalization. Nevertheless any adjustment in funding that is not incremental, that represents more abrupt change, would surely be hard to swallow in the regions that lost out. The 20 per cent of funding not allocated on a head count basis seems to us a useful beginning in terms of smoothing differences between regions but it shows that regional contracts are not *mainly* devices for regional policy in the usual sense of the term. However, and perhaps more damaging to regionalism, there can be little doubt that of the two parties coming together to agree contracts, one (the central state) is much stronger than the other (the newly set up regions). The centre therefore retains immense possibilities to award funding for those activities it particularly favours.

If regional contracts look less like instruments of regional policy in a strict sense perhaps new structures can retrieve something in terms of EU linkage? This was without any doubt a key consideration in the reform's design. Here too however problems exist. Perhaps one of the most important is lack of competent administrative capacity. In 1998 the EC rejected a swathe of Poland's PHARE applications because of quality problems (Blazyca and Kolkiewicz, 1999) and some are deeply concerned that substantial difficulties remain. As Jacek Szlachta, another respected Polish regional expert, put it in a recent round table discussion, 'At present Poland gets around €200m annually for regional development. After accession it could increase to €3bn to 6bn per annum. We are however completely unprepared to accept and utilize such large resources' (*Gazeta Wyborcza*, 2001c). John O'Rourke, from the EC's Warsaw office notes too that the actually used proportion of available PHARE funds has been falling and reached only 43 per cent in 2000, a utilization rate that threatens to undermine Poland's arguments for aid (Apanowicz, 2001a). The SAPARD programme, although set up in 2000, failed to generate any flow of resources in 2000 or even 2001. If Poland has had difficulty in absorbing transfers worth around 0.5 per cent of GDP in the very early years of this decade then how will it manage flows worth around 2.5 per cent in 2004?

Another difficulty is co-financing where, given Poland's fraught fiscal situation, meeting EU strictures is unlikely to be straightforward. Even as the first of the new regional contracts were being signed in June 2001 the respected daily newspaper *Rzeczpospolita* ran the news under the title, 'We have the contracts but what about the money' (Myczkowska, 2001a). Antoni Jankowski, the president of the Polish Association of Districts (*związek powiatów polskich*), observed in a recent discussion 'we need a solid legislative basis for regional revenue generation so that Polish regional and local authorities can be an authentic partner for Brussels' (*Rzeczpospolita*, 2001a). Matters have deteriorated sharply since mid 2001 when the finance minister, Jarosław Bauc, found that the state was facing a huge 'budget hole' (*dziura budżetowa*), that public finance was in crisis and austerity the only solution. Press reports later in the year suggested that in the face of significant real cuts in budget allocations from the centre local authorities would have to fall back on much greater commercial borrowing to fund development projects (Myczkowska, 2001b).

But for some Polish regional specialists a more fundamental deficiency is the absence of any structural or regional policy notwithstanding the 1999 reforms. One of the leading proponents of this view, Jerzy Hausner, (later to be appointed minister for labour in the SLD-UP-PSL coalition formed in October 2001) notes that 'The EU doesn't lead its members by the hand in shaping regional policy. This means that if Poland does not have its own regional policy and structures stimulating regional development even the largest sum of money will not help us' (*Gazeta Wyborcza*, 2001c). He continues, and this some two and a half years after the decentralizing reform of January 1999, 'it is disturbing we do not have our own structural policy, either in an intellectual sense or in the shape of administrative instruments thanks to which such a policy could be implemented' (*Unia&Polska*, 2001). That gloomy assessment is shared by the European Commission which noted in its 2001 annual report on Poland that in regional policy, new structures were in place that could mesh with those of the EU but despite that 'developments have largely stalled'. There was an urgent need to develop regional and central administrative capacity as well as the partnership notions on which so much of existing EU regional policy is based (Commission of the European Communities, 2001).

Table 3.1   Estimates of EU Aid Flow to Poland

Year	2000	2001	2002	2003	2004	2005	2006
€ (m)	920	920	920	3,600	4,800	6,100	7,400

Note: Assumes EU entry in 2003 and with it new sources of aid.
Source: *Gazeta Wyborcza*, 2001c.

Fears for the Future

Despite their importance, and perhaps pre-eminence, not all the anxieties regarding the future either of Polish regional policy or the regions themselves stem from internal factors. Many Polish observers are fearful that EU reforms will lead to a substantially new type of Union regional policy that Poland and Polish regions may find it difficult to connect with. Exactly how this will play out is far from clear but all interested parties are well aware of the highly politicized nature of EU regional policy, especially in the context of enlargement. Spain, the largest recipient of structural assistance in existing arrangements is not prepared to accept that it should shoulder the burden of adjustment as the EU embraces up to 12 poor new members (Crawford and Dempsey, 2001). The Spanish prime minister's argument that his country's weakest regions do not become absolutely less poor because of enlargement collides with the German view that Spain should show tangible solidarity with poor newcomers just as was shown to Spain in an earlier period. But Madrid knows that Berlin, while keen on enlargement, has its own problems with it, not least those involving labour flow, and senses that a deal can be done. Spain will support Germany's (and Austria's) desire to phase in labour mobility if Germany will ensure that Spanish regional aid is phased out in strictly manageable stages.

It is impossible to know, in late 2001, how EU regional policy will be shaped to accommodate enlargement. What is clear however is that existing rules, where the bulk of aid flows to regions where per capita GDP is below 75 per cent of the EU average (Jones, 2001), combined with an enlargement that will reduce average EU GDP per capita by around 14 per cent, generates great political tension. All 16 new Polish regions would qualify for aid post enlargement and, according to one estimate, even when broken down to (44) NUTS III type regions only Warsaw has an income level exceeding the EU average (*Rzeczpospolita*, 2001b, Bielecki, 2001c). Solutions that may be broadly acceptable involve a mix of measures like raising the income threshold to capture some at least of existing EU regions that stand to lose as well as introducing other conditions, such as persistently high unemployment, in area designation (*Economist*, 2001). But there is some fear that EU reforms may launch a much more radical departure perhaps, for example, focusing in the future on human capital and the 'information society' rather than the more traditional infrastructure investments of the past (Apanowicz, 2001b). In this vision of regional development one Polish concern is that within the country disparities will deepen as the regions likely to be more adept at hooking on to the 'infostrada' will be in the west and centre.

There is little doubt that the communications revolution offers new possibilities to shorten the distance in economic space between Polish regions but traditional communications infrastructure still remains important and is particularly poorly developed in the east. Despite many 1990s declarations on motorway construction, for example, the country had only 358km of such road in 2000 (Commission of the European Communities, 2001). Studies of the likely effect of EU membership on Polish regions suggest that the positive impact of integration will, not surprisingly, be more keenly felt in the western voivodships as well as the Katowice and Warsaw regions thus deepening regional differentiation in the future (Orłowski, 2001). There is a fear too that the EU may opt out of heavy involvement in regional policy leaving it more a matter for individual members, a 'renationalization' of regional policy as it is sometimes put. But the trend in the West is towards even deeper decentralization in regional matters, partly in response to devolution pressures and partly due to central government 'relieving itself of responsibilities' (Danson et al., 2000). One author has argued recently that a neo-liberal EU is rapidly shedding 'solidarity' and the consequence will be much greater unevenness within Europe as core and periphery differentiation becomes more pronounced in the EU while a sub-periphery, 'external' Europe is left on the EU fringe (Agnew, 2001). Others also suggest that while the European Commission asserts that cohesion and diversity go hand-in-hand, in reality the former is losing ground (Graham and Hart, 1999; Biscoe, 2001).

Conclusion

Reforms of the state administration in 1999 contributed a vital element to the decentralization promised, and started, in 1989. But more than two years later it is clear that the process is not complete. Central government structures in Warsaw have a built-in advantage in any bargaining with regions, many of which lack the deep cultural tradition, not to mention the administrative personnel and apparatus, that can act as an important generator of regional self-confidence and economic development. Lingering centralism is part of the legacy of communism. It survives not because of bad intention but due to institutional weaknesses at the local level, and, as with most post-communist transformation deficiencies, there is evidence to suggest that those weaknesses are more pronounced in the eastern parts of the country as well as in locations – such as many of the Upper Silesian towns – where state ownership remains a strong feature. The 1999 regional reforms shifted administrative boundaries but have not yet changed underlying economic realities. Genuine developments in capacity-building regional policy post 1999 have also been limited and perhaps most worrying is the gap that is opening up between a membership of the EU that may be imminent and the capacity of the regions, especially those that are weaker, to make the most of EU opportunities.

Poland expects to join the EU in 2004 and its new government, elected in September 2001, promised to accelerate the negotiating process. While there are very strong arguments in favour of 'acceleration' this should not conceal that the strategy is also risky. In the potentially short time to accession it is clear that in the field of regional policy much remains to be done. At present it seems doubtful that Polish regions, especially in the east, are ready to meet the challenges in formulating and implementing effective regional policy. Resources are scarce, perhaps scarcer than usual, and capacity limitations remain severe. The most disadvantaged regions are hit also by a general (EU and Polish) shift in policy thinking away from 'equalization' or 'solidarity' towards 'competitiveness' and 'efficiency'. This means that weaker regions in Poland and elsewhere will be at a major disadvantage within EU policy terms not only compared to peripheral regions in the EU-15 but also in relation to other better prepared accession regions in EU-27. Given this, together with the experience of previous new members on the periphery of Europe, the fear, as noted by Mike Danson in his overview of regional policy in chapter 2, that EU entry will exacerbate existing regional differences should not be too swiftly dismissed.

Appendix

Table 3.2   Poland's Post 1999 Regions, Salient Features

Region	GDP/capita, 1998 Poland = 100	GDP/capita, 1998		Unemployment 1999 2000	
		EU15 =100	EU27 =100	(% rate, end year)	
Mazowieckie	146.1	53	61	9.1	12.5
Sląskie	111.9	40	47	9.7	16.6
Wielkopolskie	105.6	38	44	9.8	12.0
Dolnosląskie	99.8	36	42	13.8	19.3
Pomorskie	98.7	36	41	13.0	19.0
Zachodnio-pomorskie	97.7	35	41	14.9	21.7
Kujawsko-pomorskie	92.2	33	38	15.2	19.7
Lubuskie	91.2	33	38	15.3	20.5
Małopolskie	91.0	33	38	9.7	12.6
Łódżkie	88.6	32	37	12.7	15.3
Opolskie	88.3	32	37	12.1	14.8
Swiętokrzyskie	77.2	28	32	15.6	19.6
Warmińsko-mazurskie	76.7	28	32	21.3	24.4
Podlaskie	76.3	28	32	11.4	15.6
Podkarpackie	75.9	27	32	16.1	20.1
Lubelskie	72.5	26	30	12.6	14.4
POLAND	100.0	36	42	12.3	16.3

Notes: GDP data expressed in purchasing power standards (PPS) as compiled by Eurostat and unemployment rates, also compiled by Eurostat follow ILO standard definitions.

Sources: Rocznik Statystyczny Województw, 2000, Główny Urząd Statystyczny, Warszawa, 2000; EU comparative GDP data from *Eurostat, News Release* No. 31, 15 March 2001 and unemployment data from Behrens, 2001.

References

Agnew, J (2001) 'How Many Europes?: The European Union, Eastward Enlargement and Uneven Development' in *European Urban and Regional Studies*, Vol. 8. Issue 1, January.

Apanowicz, P. (2001a) 'Co wspólnego ma Mazowsze z Andaluzją', *Rzeczpospolita*, June 18.

Apanowicz, P. (2001b) 'Zaproszenie do dyskusji', *Rzeczpospolita*, May 5-6.

Behrems, A. (2001) 'Regional unemployment rates in the Central European Candidate Countries 2000', *Statistics in Focus: General Statistics*, Theme 1- 8/2001, Eurostat, European Communities.

Bielecki, J. (2001a) 'Bruksela aprobuje program rządu', *Rzeczpospolita* July 17.

Bielecki, J. (2001b) 'Inwestorzy jednak nie stracą', *Rzeczpospolita*, October 19.

Bielecki, J. (2001c) 'Całej Polsce przysługuje pomoc', *Rzeczpospolita*, April 9.

Biscoe, A. (2001) 'European integration and the maintenance of regional cultural diversity: symbiosis or symbolism?', *Regional Studies* ,33.

Blazyca, G. & Kolkiewicz, M. (1999) 'Poland and the EU: Internal Disputes, Domestic Politics and Accession', *The Journal of Communist Studies and Transition Politics*, Vol. 15, December, No. 4.

Buxton, J. (2000) 'Business births falter north of the border', *Financial Times*, February 24.

Commission of the European Communities (2001) *Regular Report on Poland's Progress Towards Accession*, Brussels, November 13.

Crawford, L. and Dempsey, J. (2001) 'Aznar digs in for a fight', *Financial Times*, May 17.

Danson, M. Halkier, H. and Cameron, G. 'Regional governance, Institutional Change and Regional Development' in *Governance, Institutional Change and Regional Development*, Mike Danson, Henrik Halkier and Greta Cameron eds, Ashgate, Aldershot.

Dziemianowicz, W. (1999), 'Dynamika międzyregionalnych zróżnicowań społeczno-gospodarczych w Polsce w latach 1990-98' in *Rozwój regionalny jako element strategii społeczno-gospodarczej Polski w latach 2000-2006*, Jacek Szlachta, Andrzej Pyszkowski eds, Instytut Badań nad Gospodarką Rynkową, Gdańsk.

Dziennik Ustaw (2000a) 'Ustawa z dnia 12 maja 2000r. o zasadach wspierania rozwoju regionalnego', Nr 48, poz. 550, June 14.

Dziennik Ustaw (2000b) 'Rozporządzenie rady ministrów z dnia 28 grudnia 2000r. w sprawie przyjęcia Programu wsparcia na lata 2001-2002', Nr 122, poz. 1326, December 31.

Economist (2001) 'The EU's regional aid: What's ours is ours', May 26.

Gazeta Wyborcza (1999) 'Polska A, B i C' (Poland A, B and C), August 4.

Gazeta Wyborcza (2001a) 'Reforma górnictwa – pakiet stracił impet', May 10.

Gazeta Wyborcza (2001b) 'Nie tylko zachód', June 21.

Gazeta Wyborcza (2001c) 'Warszawa trzyma kasę', June 19.

Gazeta Wyborcza (2001d) 'Trzeba umieć brać', March 22.

Gilowska, Z. (2000), 'Regionalne uwarunkowania reform strukturalnych' in *Studia Regionalne i Lokalne*, Vol 2, No. 2.

Golinowska, S. (1998) 'Zróżnicowanie regionalne a procesy migracyjne' in *Rozwój ekonomiczny regionów. Rynek pracy. Procesy migracyjne - Polska, Czechy, Niemcy*, Golinowska, S. (ed.), Instytut Pracy i Spraw Socjalnych, Warszawa.

Gorzelak, G. (1996) *The Regional Dimension of Transformation in Central Europe*, Jessica Kingsley Publishers for the Regional Studies Association, London.

Gorzelak, G. (1998) *Regional and Local Potential for Transformation in Poland*, European Institute for Regional and Local Development, Warsaw, Paper 14.

Gorzelak, G., Jalowiecki ,B., Herbst M. and Roszkowski,W. (1999) '*Transformacja systemowa z perspektywy Dzierzgonia*', European Institute for Regional and Local Development, Scholar, Warszawa.

Gorzelak, G. (2001), 'Decentralization, regional development and regional policies' in *Poland into the New Millennium*, George Blazyca and Ryszard Rapacki eds, Edward Elgar, Cheltenham.

Graham, B. and Hart, M. (1999) 'Cohesion and diversity in the European Union: irreconcilable forces?', *Regional Studies*, 33.

Hausner, J. Kudłacz, T. and Szlachta, J. (1995) *Regional and Local Factors in the Restructuring of Poland's Economy. The Case of South-East rn Poland*, Kraków Academy of Economics,UCEMET, Kraków.

Heffner, K. (1998) 'Entwicklung und Zusammenarbeit im deutsch-polnischen Grenzraum' in *Grenzübergreifende Kooperation in östlichen Mitteleuropa. Beiträge zu einem politik-und regionalwissenschaftlichen*, Symposium an der TU Chemnitz, Neuss, B., Jurczek, P. and Hilz, W. (eds), Europäisches Zentrum für Föderalismus Forschung, Tübingen.

Heffner, K. and Solga, B. (1999) *Praca w RFN i migracje polsko-niemieckię a rozwój regionalny Śląska Opolskiego*, Państwowy Instytut Naukowy, Instytut Śląski w Opolu, Opole.

Hunek, T. (2001) 'Rolnictwo w procesie formowania się nowego ładu społeczno-gospodarczego Polski' in *Wieś i rolnictwo na przełomie wieków*, Bukraba-Rylska, I. and Rosner, A. (eds), Instytut Rozwoju Wsi i Rolnictwa PAN, Warszawa.

Jensen, C. and Leijon, S. (2000) 'Persuasive Storytelling about the Reform Process: The case of the West Sweden Region' in *Governance, Institutional Change and Regional Development*, Mike Danson, Henrik Halkier and Greta Cameron eds, Ashgate, Aldershot.

Jones, R. A. (2001) *The Politics and Economics of the European Union*, Edward Elgar, Cheltenham.

Kafkalas, G. and Thoidou, E. (2000) 'Cohesion Policy and the Role of RDAs in the Making of an Intelligent Region: Lessons from the Southern European Periphery' ' in *Governance, Institutional Change and Regional Development*, Mike Danson, Henrik Halkier and Greta Cameron eds, Ashgate, Aldershot.

Kekic, L. (2000) 'EU enlargement: what to expect?' in *Economies in transition: Regional Overview*, EIU Ltd, London.

Kryńska, E. (2000) *'Polskie specjalne strefy ekonomiczne'*, European Institute for Regional and Local Development, Warszawa.

Kuźnik, F. (2000) 'Regionalna struktura, skala i specyfika procesów restrukturyzacji przemysłu z określeniem obszarów problemowych z punktu widzenia strategii rozwoju regionalnego w latach 2000-2006', in *Narodowa strategia rozwoju regionalnego*, Szlachta, J. (ed.), Warszawa.

OECD (1992) Regional Development Problems and Policies in Poland, Paris.

OECD (1996) Transition at the local level. The Czech Republic, Hungary, Poland and the Slovak Republic, Paris.

Myczkowska, A. (2001a), 'Kontrakty są, ale czy będą pieniądze', *Rzeczpospolita*, June 30 - July 1.

Myczkowska, A. (2001b) 'Mniej na dotacje i kontrakty wojwodzkie', *Rzeczpospolita*, November 28.

Oktaba, L. (2001) 'Ta pomoc się połaca', *Rzeczpospolita*, April, 19.

Orłowski, W.M. (2001) 'Polskie regiony na tle wyzwań integracyjnych' in *Polityka regionalna państwa polśród uwikłań instytuconalno-regulacyjnych*, Szomburg J., ed, Instytut Badań nad Gospodarką Rynkową, Gdańsk.

Pisz, Z. (2001) *Problemy społeczne transformacji*, Akademia Ekonomiczna im. Oskara Langego we Wrocławiu, Opole.

Rocznik Statystyczny (1996, 2000), GUS, Warsaw.

Rocznik Statystyczne Województw (1997, 2000), GUS, Warsaw.

Rosner, A. (2001) 'Społeczno-ekonomiczne uwarunkowania przemian strukturalnych w rolnictwie',in *Wieś i rolnictwo na przełomie wieków*, Bukraca-Rylska I. and Rosner A. (eds), Instytut Rozwoju Wsi i Rolnictwa PAN, Warszawa.

Rykiel, Z. (1995) 'Polish core and periphery: under economic transformation', in *Geographia Polonica*, vol. 66.

Rzeczpospolita (1998) 'Zachód ucieka na Zachód, wschód na Wschód', April 1.

Rzeczpospolita (2001a) 'Polska w Europie regionów', July 25.

Rzeczpospolita (2001b) 'Najważniejszy jest wpływ na decyzje', March 7.

Scottish Enterprise (1993) Scotland's *Business Birth Rate: A National Enquiry*, Scottish Enterprise, Glasgow.

Scottish Enterprise (2001) *The Review of the Business Birth Rate Strategy*, Scottish Enterprise, Glasgow.

Stenning, A. (2000), 'Placing (Post-) Socialism: The Making and Remaking of Nowa Huta, Poland' in *European Urban and Regional Studies*, Vol. 7, Issue 2, January.

Stryjakiewicz, T. (1996) 'Uwarunkowania polsko-niemieckiej współpracy przygranicznej na tle polityki regionalne' in *Problemy współpracy regionalnej w polsko-niemieckim obszarze przygranicznym*, Chojnicki   Z. and Stryjakiewicz, T. eds Komietet Przestrennego Zagospodarowania Kraju PAN, Biuletyn KPZK PAN, No. 171, Warsaw.

Surażska, W. (1999) 'Gdzie Polska rozwija się najszybciej', *Rzeczpospolita*, September 17.

Swianiewicz, P. and Dziemianowicz, W. (1999) 'Najlepiej być blisko metropolii', *Rzeczpospolita*, February 11.

Unia&Polska (2201), 'Szansa może być zagrożeniem', 30 July-13 August.

Web sites (with some English language pages):
Chancellery of the prime minister: www.kprm.gov.pl.
Economics ministry: www.mg.gov.pl.
Committee for European Integration: www.kie.gov.pl.
European Institute for Local and Regional Development, University of Warsaw: www.eurreg.uw.edu.pl.

Chapter 4

Polish Regional Policy and the Problems of Upper Silesia After the First Transformation Decade

Andrzej Klasik and Krystian Heffner

The Influence of the Past

Diversity across Polish regions in economic development is highly dependent on the past as well as on the economic structure of regions.

First, the socio-economic history of the last 200 years has had a profound impact on regional development in Poland. Even today differences between former Russian, Austro-Hungarian and Prussian partitions and annexations remain clearly visible. Of course, nearly half a century of communist economy has also had an impact. Second, during the socialist era, regional investment location was not the outcome of market forces but due to administrative decisions. The difference between the more industrialised western Poland (so-called 'Poland A') and the more agricultural eastern Poland ('Poland B') survived almost intact during the entire socialist period (see Kozak, M. and Pyszkowski, A., 2000, pp. 32-40; Golinowska, S., 2001a, pp. 83-121). The third important factor is the impact of transformation in the 1990s. It is commonly held that transition promoted strong regional polarisation giving bonuses for more urbanised Western regions. Matters however are more complex. In the early 1990s regional differentiation decreased (as measured, for example, by coefficients of variation in regional GDP per capita), the result of the dramatic decline in industrial production. From then regional diversity slowly increased. In that period it was not industrial production that mattered but rather the development of market services. In the 1990s, changes in the Polish economy's spatial structure and in regional diversity were much more modest, especially if Warsaw is taken out of the picture. But even if there was no dramatic polarisation, greater regional diversity was evident in statistics on regional GDP per capita, more urbanised regions doing better and the rural regions less well (Żienkowski, 1997, pp. 7-32). Generally speaking economic diversity across Polish regions is substantial but not dramatically high.

Two regions, Upper Silesia and Łódź voivodships, have an unusually strong concentration of industry. Previously, industry was their most important development asset even if it was associated with environmental problems. On the other hand, eastern and north-eastern regions are relatively strongly dependent on agriculture. Of course this does not mean that agriculture had to be a drag on development – as Wielkopolska and Opole-Silesia show. Rather, what is important is the type and effectiveness of regional agriculture as well as the existence of a modern food processing industry. Other regions, particularly Warsaw, Gdańsk and Szczecin, have an exceptionally strong share of services in GDP. If Warsaw clearly benefited during the 1990s, the scene was more complicated in both Baltic coastal regions. There, services did not develop as fast as in Warsaw but the economic situation in both regions is much better than in those strongly dependent on industry or agriculture (Orłowski, 2001, pp. 52-88).

Diversity and Development

Diversity stems from the natural features of regions as well as socio-economic factors. The latter however are the most important. Administrative reform, as we will see below, also plays a part. Diversity is evident in indicators of socio-demographic potential[1], infrastructure, the level of industrial development, the role of agriculture, environmental conditions as well as research and scientific potential.[2]

Following the reform of state administration the former 49 voivodships were, as noted in chapter three, merged into 16 new regions from 1999. As might be expected, the larger regions have very wide intra-regional differences in development level. Despite internal polarisation (for example between Warsaw and Radom in Mazowieckie or Poznań and Konin in Wielkopolskie) regional imbalances in Poland are not dramatically high by EU standards. The Warsaw region is an unquestioned leader but, it apart, inter-regional diversity in Poland is similar to that found in EU countries (Kudlacz, 2001, pp. 107-120). One important, although in a sense artificial, factor acting to diminish Polish regional diversity is administrative reform. The larger regions created from 1999 reduced inter-regional diversity but promoted intra-regional differences. Polish regions can be classified into six types (Kudlacz, 2001, pp.16-51):[3]

1. The very special Warsaw region, Mazowsze, which is dissimilar to all the others because of its exceptionally high level of development across almost all dimensions (except agriculture) and its huge research and scientific potential. But Mazowsze is an administrative unit composed of highly developed Warsaw with surrounding agglomeration and a large but weakly developed and problem 'interior'.

2. Silesia with its well-known features. Silesia has great development problems because of the sheer concentration of 'sunset' sectors. In the early 1990s there

---

[1] Socio-demographic potential reflects the balance of the natural increase of population, internal and foreign migration, age and dependency relationships, participation rates, productive age structure, population with university level education and so on. See for example Kołodziejczyk, D, et al, (1998); Pisz Z, (2001).

[2] The research and cientific development potential refers to student, professors and scientist numbers, higher education institutions, patents and such like.

[3] For other approaches to typecasting regions see Gorzelak (2001), pp. 209-210.

was a great fear that the region would witness large-scale labour shedding and inward investor hesitance, not least because of environmental issues. Facing a sudden collapse in traditional markets for coal and steel Silesia was widely thought to be an obvious potential transformation flashpoint (Gorzelak, 1998). It is rightly regarded as a special regional development case because of its exceptional concentration of heavy industry as well as its particular infra-structural, environmental, labour market and social problems. Polish researchers have observed that the heavily industrialised Silesian urban centres, despite populations with higher than average family incomes, display little enterprise, 'a fundamental problem closely associated with mining restructuring' (Surażska, 1999). The Silesian regional problem has clearly not disappeared during twelve years of transition and still needs an active transformation policy.

3. Regions with a better demographic potential and a modern economic structure, that is, a high share of employment in services. Three of the post 1998 voivodships, from the northern and north-western part of Poland (Zachodniopomorskie, Pomorskie and Lubuskie) fall into this category.

4. This group (Dolnosląskie centred around Wrocław and Małopolska around Kraków) also has a higher level of demographic potential and a modern economic structure but other development factors tend to the average, particularly concerning environmental problems, R&D potential and agricultural development.

5. Regions with relatively highly developed agriculture with good natural environment. Four voivodships in mid and southern Poland can be included: Wielkopolska (with Poznań its centre), Łódź, Opole and Kujawsko-Pomorskie (Toruń-Bydgoszcz).

6. The final group is characterised by a low level of development (especially in industrial development and the research and scientific base) but has the highest quality natural environment. Five voivodships may be included in this group: Podlaskie (Białystok), Podkarpackie (Rzeszów), Warmińsko-Mazurskie (Olsztyn) and Świętokrzyskie (Kielce).

Groups three to five are fairly uniform with the exception of the regions with big urban agglomerations (Wielkopolska, Łódź, Lower Silesia and Małopolska) and are relatively undifferentiated even from each other. On the other hand the first, second and sixth groups are qualitatively different and cannot be simply compared to other Polish regions.

As noted earlier, the substantial enlargement of regions from 1999 is associated with great intra-regional diversity in development and living conditions. Mazowieckie, for example, has extremely high internal economic diversity as a result of different development paths in the very special Warsaw agglomeration as compared to the region's more peripheral parts. Other big regions (Podlaskie, Podkarpackie, Łódź and Małopolska) also have relatively great internal differentiation. The least intra-regional development diversity is found in Lubuskie and Warmińsko-Mazurskie, regions without any big urban agglomeration.

During the 1990s, the most important shift in regional development shows that Mazowieckie with Warsaw became a lonely leader, its growth much faster than all other regions. One other, but still much weaker leader in development was, perhaps surprisingly, Silesia. Wielkopolskie region was another. Despite its relative success in the 1990s industrial centres like Silesia are in grave danger. The group of lagging regions (Lubelskie, Podlaskie, Podkarpackie, Warmińsko-Mazurskie and Świętokrzyskie) remains however fairly constant, the distance separating them from more developed regions widening. In the fourth and fifth groups above change has led in some cases to relative success (Wielkopolskie, Małopolskie and Łódzkie) and failure (Opolskie, Kujawsko-Pomorskie).

The most important 1990s changes in regional diversity were in private sector developments, its share in employment and industrial production. Other dimensions (for example, GDP per capita and average earnings) also showed significant changes across regions. Regions with relatively more agriculture and less industry and services were weaker and performed less well than others. With hindsight those structural features should have been reflected in national regional policy.

Generally, fundamental changes in regional diversity did not appear during the 1990s despite the fact that numerous researchers and politicians have suggested so, usually however based on single factors such as the share of market services, average earnings, unemployment and so on.

Regional Effects of Polish Accession to the European Union

Polish regional policy in the pre-accession period is determined by both external and internal factors. Four aspects seem particularly important (Klasik, 2001, pp. 214-222). First, the changing social, economic and environmental development level in the regions themselves. Second, the institutional framework of regional policy, the state of public finances and the likely dynamic of economic growth in Poland. Third, Polish-EU relations and joint developments in regional policy, particularly regarding structural and cohesion funds. Fourth, the international economic situation including information society developments, progress in science and technology and globalisation of markets.

From the EU viewpoint all Polish voivodships are lagging regions and qualify under existing rules for structural fund support. Many Polish and foreign experts have shown that the competitiveness of Polish regions is generally low and restructuring progress is limited (Winiarski, 2000, pp. 11-19).

Looking first at the spread of living standards the highest incomes per capita in 1999 were found in Mazowieckie and Śląskie voivodships, the lowest in Podkarpackie, Warmińsko-Mazurskie and Lubelskie.

Rural regions, along with urban regions with a very high concentration of declining industries, are in the most difficult position. It seems realistic to assume that in the first phase of EU membership support for regional development will be limited. In the period 2000-2006 Polish regional policy should, we believe, have two principal aims. It should strengthen regional competitiveness and counteract marginalisation. Looking further ahead (to 2012) a longer term regional policy should shape innovative regional structures aiming for an increasing share of service and R&D in regional economic structure and transformation of depressed regions by acceleration of heavy industry privatisation, FDI attraction and greater social mobility. Generally, Polish regions can be divided to two groups needing two types of regional policy at the national level. First, a pro-competitive regional policy for regions with big agglomerations and effective urban centre and second a redistributive regional policy aimed at counteracting marginalisation, supporting conversion of traditional and depressed economic structures and concentrating support on agriculture.

Table 4.1 GDP in Poland by Voivodships in 1995 and 1999

Voivodship	1995	1999
	GDP *per capita* (zł)	
Mazowieckie	9,941	23,760
Śląskie	9.824	17,565
Wielkopolskie	7,859	16,747
Dolnośląskie	8,298	16,273
Pomorskie	8,012	16,120
Zachodniopomorskie	8,217	15,924
Łódzkie	7,289	14,497
Lubuskie	7,850	14,444
Małopolskie	7,066	14,231
Kujawsko-Pomorskie	7,951	14,121
Opolskie	7,877	13,320
Świętokrzyskie	6,341	12,435
Warmińsko-Mazurskie	6,365	12,341
Podkarpackie	6,183	11,685
Podlaskie	6,053	11,580
Lubelskie	6,153	11,112
Range: highest-lowest	*3,888*	*12,648*
Poland	7,985	14,316

Source: own calculations based on Statistical Yearbooks, GUS Warszawa.

It is widely accepted that EU membership should increase welfare and promote economic growth but it is not out of the question that the accumulation of adjustment costs in some regions can lead to losses even in other regions. EU accession will accelerate industrial restructuring through stiffer competition and the elimination of any final trade barriers. Very high adjustment costs will hit, in particular, such protected sectors as coal, steel, petrochemicals and food industries. Meanwhile it can be expected that the services sector will expand rapidly, perhaps enjoying a boom. Agriculture and rural areas will also experience fierce restructuring. Tensions on regional labour markets will very clearly appear. With this general background we turn below to a more detailed description of the Silesian regional economy.

Table 4.2　Problem areas for Polish regional policy

Identification criteria	Voivodships and regional capitals
Barriers in making use of endogenous development potential	

- technical infrastructure
- telecommunication
- business environment and enterprise

- university level education
- environmental conditions
- tax incomes

- Podlaskie, Warmińsko-Mazurskie, Lubelskie
- Podkarpackie, Świętokrzyskie, Śląskie
- Opolskie, Świętokrzyskie, Podlaskie, Warmińsko-Mazurskie, Lubuskie,
- Podkarpackie, Świętokrzyskie, Lubuskie
- Śląskie
- Podlaskie, Podkarpackie, Lubelskie, Świętokrzyskie, Warmińsko-Mazurskie,

Weakness and lack of metropolitan functions

- services
- university level education
- international business

- hotels and tourism
- telecommunication

- Upper-Silesian agglomeration, Łódź, Bydgoszcz
- Białystok
- Bydgoszcz, Rzeszów, Olsztyn, aglomeracja górnośląska,
- Białystok, Olsztyn, Kielce, Opole, Lublin
- Bydgoszcz, Zielona Góra, aglomeracja górnośląska, Łódź,
- Opole, Bydgoszcz, Kielce, aglomeracja górnośląska

Relatively low level of development and structural transformation

- low economic development
- structural unemployment
- industrial restructuring
- agriculture restructuring and low population density

- Lubelskie, Podlaskie, Opolskie
- Świętokrzyskie, Warmińsko-Mazurskie, Kujawsko-Pomorskie, Pomorskie, Zachodniopomorskie
- Śląskie
- Lubelskie, Podlaskie

Sources: Polska w nowym podziale terytorialnym, GUS, Warszawa 1998; Orłowski, et al (1998); Dutkowski, M, and Gawlikowska - Hueckel M, (1998).

Silesia

General Demographic and Socio-economic Characteristics

The historically strong heavy industry base in Silesia developed further during socialism when coal and steel were regarded as critical for the construction of socialism. The Silesian working class was privileged economically (in wages, early pensions, bonuses, flats, holidays and easier access to consumer goods), and to some extent socially, earning high prestige among other groups of Polish society. In the 1990s, however, coal and steel became a heavy burden, sooner or later to face restructuring. Yesterday's avant-garde now appeared as a loser in systemic transformation.

The Silesian voivodship (*województwo śląskie*) in its post 1999 structure has 3.9 per cent of Poland's land area and, with 4.9m people, 12.6 per cent of the population. It is the fourteenth region in Poland by area but the second largest in terms of population. giving it the highest population density in Poland (397 inhabitants/km^2), more than three times the national average (124 inhabitants/km^2)[4]. Administrative reform gave the new Silesian voivodship 86 per cent of the area of the former Katowice administrative region plus 70 per cent of former Częstochowa and 60 per cent of Bielsko-Biala voivodships. The new region has 19 towns with district (*powiat*) status and 17 rural districts. There are 166 municipalities (*gmina*).

From a demographic viewpoint Silesia, besides Mazowsze and Łódź, has the lowest natural increase-rate of population in Poland (minus 0.5 per 1,000 inhabitants in 2000 compared to plus 0.2 in Poland), the result of minus 0.8 per 1,000 in urban, and plus 0.5 in rural areas. The region is losing population with, in 1999, a net outflow of 1.5 people per 1,000 inhabitants. Infant mortality, at 10.6 per 1,000 live births remains high compared to the national average of 8.9. Silesia has the highest urbanisation rate in Poland, with almost 80 per cent (3.9m) of the population living in towns. Of the region's 1.6m dwellings almost 16 per cent is municipal property, the second largest proportion in Poland. Population density is highly differentiated within the region, highest in the cities in Upper-Silesian agglomeration (Świętochłowice has 4,558 inhabitants/km^2 and Chorzów 3,580) and lowest in the north-eastern part of region (Żarnowiec with only 40). On average the Silesian population is younger than in other regions, its 62.5 per cent in the working age category compares to 60 per cent as a national average. Nevertheless, the process of ageing is faster than in other regions. Generally, socio-demographic characteristics differ between cities and rural areas (for example the number of divorces in cities is 80 per cent higher than in rural areas, the number of live births is 20 per cent higher in rural areas).

Some 6.8 per cent of the Silesian population is in higher education and only 3.6 per cent have incomplete primary or no education. Silesia, with 140,000 students in higher education is second in absolute terms only to Mazowsze with its 253,100 students and comes ahead of Małopolska (117,900) and Lower Silesia (102,800). Despite the large number of students in total the participation rate in higher education per 1,000 inhabitants is lower than national average (32.0 and 36.8 respectively). The region has a high percentage of its youth (69 per cent) studying in technical schools (including vocational schools, mainly private in Katowice, Tychy, Bielsko-Biała, Częstochowa, Ruda Śląska). A declining interest in vocational education that prepares students directly for specific types of work in favour of general education is a positive tendency.

In 1995, Silesia produced 15.7 per cent of Poland's GDP, and at 9,800 zł per capita it was 23 per cent above the national average. By 1999 it was producing 13.9 per cent of the total and at, 17,600 zł per capita, was only 10 per cent more than the national average. The industrial profile of Silesia is reflected in the large number (805,800) employed in industrial and construction. The employment index in those sectors, at 165 per 1,000 inhabitants, is higher than national average. The number

---

[4] Further details can be found in *Development Strategy of the Silesian Voivodeship 2000-20*, Urząd Marszałkowski w Katowicach, Katowice 2001.

employed in market and non-market services is about 895,400 (183.4 per 1,000 inhabitants). In 1999, Silesia had some 226,600 SMEs, of which 3,500 had foreign capital participation. Silesia has 110,00 private farms, mostly small (80.6 per cent have between 1 and 5 ha of arable land). In 1997 some 338,200 people earned a living, exclusively, or on a dual profession basis, from farming. Farming is predominantly concentrated in the northern (around Częstochowa) and southern (Bielsko-Biała) parts of the region. A general profile of Silesia is summarised in Appendix table 4.4.

Table 4.3 Registered Unemployment in 2001 (end June)

Voivodships	Registered Unemployed Persons in 2001						
					% of total		
	('000)	(% rate)	June 2000 = 100	women	unemployed for longer than 1 year	With higher education	up to 25 years old
Warmińsko-	165.1	25.5	114.9	54.6	43.0	1.9	38.0
Lubuskie	94.6	21.6	119.4	53.3	50.2	1.9	26.8
Zachodniopomorskie	156.8	21.4	121.7	54.3	46.6	2.4	29.3
Kujawsko-pomorskie	190.0	20.2	114.1	56.4	50.4	1.8	32.9
Dolnośląskie	243.7	19.0	117.0	54.2	47.9	2.2	30.2
Pomorskie	157.1	17.2	121.0	56.1	41.2	2.5	31.0
Świętokrzyskie	122.8	17.1	111.9	52.1	49.4	3.8	26.6
Łódzkie	214.3	16.6	110.8	51.4	46.1	2.8	33.9
Podkarpackie	183.2	16.1	110.9	53.1	47.9	2.8	27.9
Opolskie	71.8	15.9	117.3	56.8	53.0	2.6	31.7
Śląskie	290.2	14.2	125.1	56.5	51.3	2.7	29.0
Lubelskie	162.8	14.2	109.8	51.4	43.0	3.7	27.2
Podlaskie	79.2	13.7	110.2	51.8	42.8	3.3	27.7
Wielkopolskie	208.3	13.2	121.8	56.3	44.5	2.0	25.2
Małopolskie	196.0	12.6	118.8	54.6	48.5	2.9	27.2
Mazowieckie	313.3	11.9	119.0	52.2	45.0	2.9	26.9
Polska	**2849.2**	**15.8**	**116.9**	**54.1**	**43.3**	**2.6**	**27.1**

Source: Information on socio-economic situation of voivodships, Central Statistical Office, Warsaw No. 2/2001.

The Labour Market

In 1999, Silesia had a labour force of some 2.1m with 1.9m in employment and 210,300 unemployed. But the Silesian labour market became unstable at the turn of the century and unemployment grew to 290,200 by mid 2001. Industry and construction sectors account for 42.1 per cent of total employment compared to a national rate of 28.8 per cent. Non-market services had 14.3 per cent and agriculture 11.1 per cent. The activity rate (a ratio of the number of employed persons to total population) is 38.9 per cent. Despite the restructuring process, steel, power generating, machinery, automotive and mining companies still play a dominant role in the labour market, especially in the Katowice agglomeration.

The unemployment rate, at 14.2 per cent, was lower than the national average (15.8 per cent) in mid 2001 but the situation is deteriorating (in 1998 the unemployment rates were 9.9 per cent and 13 per cent respectively). The region had the fastest increase in unemployment in Poland in 2000-2001. Moreover, that tendency looks pretty permanent because restructuring in steel and mining and in the public service sector (education, health service and railways) is still far from complete. The Silesian region, from the viewpoint of deepening labour market disequilibrium has become one of the most troublesome and problematic areas in Poland.

Among the unemployed in Silesia 32.2 per cent are aged 18-24 years, 24.9 per cent 25-34 years, 25.6 per cent in the range 35-44 years and 12.3 per cent over 44 years of age. Most of the unemployed have poor education. Some 32.4 per cent have only basic or incomplete education and 38 per cent have vocational education only. The situation of women in the labour market is less favourable than that of men. Some 59 per cent of registered unemployed are female. In the past the largest proportion of the Silesian unemployed (21.6 per cent) found a new job between 6 and 12 months of unemployment with 19.7 per cent found a job in 3-6 months but the number in long term unemployment is constantly increasing. Only 17.8 per cent of the unemployed were entitled to benefits and for those living in rural areas that proportion fell to 14.5 per cent.

Principal Development Problems in Silesia

Upper Silesia, in its new administrative structure, needs a regional policy that is sensitive to its deep restructuring problems and its particular social and labour tensions. But it must be a regional policy tuned to the general framework of Polish and future EU regional policies. Most development problems in Silesia are connected to its highly differentiated internal structure, clearly visible at the NUTS3 level. The northern part of the region (Częstochowa) is characterised by weak infrastructure, poor and archaic agriculture and a generally low level of education. The central part's (the Upper Silesian Agglomeration) traditional heavy industrial economic structure acts as brake on growth. Problems here include: delays in restructuring steel, mining and power industries; slow privatisation; weak absorption of research and innovation by companies; deep labour market imbalances characterised by a poor match between education profile and changing labour market conditions; greatly depreciated housing and huge environmental hazards diminishing the quality of life; the underdevelopment of the transport infrastructure with, in particular, congestion in the East-West communication axis; still unsolved problems with mining industry waste and a concentration of degraded post-industrial areas. The south-western part of the region (the Rybnik coal mining district) has also been badly hit by difficulties caused by delays in restructuring of coal mining sector, weak infrastructure and communication systems. In the southern part (Bielsko-Biała) tensions are fuelled by fast unemployment growth, poor infrastructure at border crossing points and still underdeveloped trans-border co-operation with the Ostrava region in the Czech Republic. There are also problems in environmental protection and the sustainable use of forests and rivers for tourism and recreation.

Yet all is not gloomy and it is curious that in comparable external and internal socio-economic conditions some Silesian towns have developed dynamically (Katowice, Bielsko-Biała, Gliwice, Tychy, Żywiec) while other are in what seems to be permanent economic stagnation (Racibórz, Jastrzębie Zdrój, Dąbrowa Górnicza, Świętochłowice, Siemianowice, Będzin).

Conclusions

After more than a decade of profound economic and social reforms Silesia is still struggling to find its place in the new Poland and in Polish regional policy. Before 1989 Silesia, with the highest concentration in Poland of heavy industry, its strong trade unions and its developed urbanisation was the leader among Polish regions. Since 1989 the most important change for Silesia is probably its fall from that number one position.

Silesia, in the nearest future, must complete the restructuring and privatisation of traditional sectors. It needs also to counter its negative image in Poland and abroad as an area of environmental hazard with a crisis-generating economic structure.[5]

The challenge for a regional policy for Silesia is how to match the highly competitive position of Warsaw and regional centres like Kraków, Poznań and Wrocław and their successes in higher education, research and development, foreign capital inflow, international co-operation and winning EU funding. Many formerly heavy industrial regions in western Europe have undergone, it seems, a more or less successful restructuring. The Ruhr is one example and of course the West of Scotland another. While lessons can undoubtedly be drawn from those (sometimes traumatic) cases it is worth emphasising the uniqueness of the Upper Silesian experience where restructuring is taking place alongside a simultaneous economic and social transformation of the entire country and EU accession not to mention a permanent lack of financial resources.

---

[5] For more details see *Development Strategy of the Silesian Voivodship 2000-20*, Urząd Marszałkowski w Katowicach, Katowice 2001.

Appendix

Table 4.4 A General Profile of the Silesian Voivodship

Population	4.8 m (second place after Mazowieckie)
Area	12,300 km² (14th place)
Population density	397 inhabitants/km² (3.2 times the national average)
Administrative structure	19 urban districts, 17 rural districts, 166 municipalities
Urbanisation	79.6% (highest in Poland)
Dwellings inhabited	1.6m, of which 15.7% are municipal (13.6% of total number of dwellings in Poland)
Higher education	140,000 students, 28.7 students per 1,000 inhabitants
Gross Domestic Product (GDP) per capita	14.6% of the Polish total, zł 3,300 per capita (15% above the national average)
Number of businesses of which: companies with international capital sole traders	353,000 (12.6% of the total number registered in Poland) 3,100 (0.9% of the total number registered in the voivodship) 281,300 (79.7% of the total registered)
Population over the age of 15 earning a living from agriculture	338,200 (1.3% of the total earning a living from agriculture in Poland)
Employment in industry and construction	805,800 (42.1% of total employment in the voivodship)
Employment in services sector	895,400 (46.8% of total employment in the voivodship)
Unemployment rate	14.2% (national average 15.8% in 2001)

Source: Regional Development Strategy for Śląskie Vovodship, Katowice 2001.

References

Dutkowski, M. (2001), *'Typologia polskich regionów'*, (A typology of Polish regions) in J. Szomburg (ed), *Polityka regionalna państwa pośród uwikłań instytucjonalno-regulacyjnych*, (The state's regional policy in institutional-regulatory context), Instytut Badań nad Gospodarką Rynkową, Gdańsk.

Dutkowski, M. and Gawlikowska - Hueckel.L, (1998), *Sytuacja społeczno-gospodarcza nowych województw*, (The socio-economic situation of the new voivodships), Instytut Badań nad Gospodarką Rynkową, Gdańsk.

Kozak, M, and Pyszkowski, A, (2000), 'Uwarunkowania rozwoju regionalnego Polski i podstawowe dylematy polskiej polityki regionalnej' (Conditions of regional development of Poland and basic dilemmas of Polish regional policy); in *Polityka regionalna i jej rola w podnoszeniu konkurencyjnbości regionów*, Klamut M. and Cybulski L., eds, Wydawnictwo Akademii Ekonomicznej we Wrocławiu.

Golinowska, S, (2001a) 'Economic Development and the Labour Market in Poland 1989-1996' in *Economic and Labour Market Development and International Migration - Czech Republic, Poland Germany*, Hönekopp, E., Golinowska, S., Horalek M., eds, Bundesanstalt für Arbeit, Nürnberg.

Gorzelak, G, (1998), *Regional and Local Potential for Transformation in Poland*, European Institute for Regional and Local Development, Warsaw 1998, Paper 14.

Gorzelak, G, (2001), 'Decentralisation, regional development and regional policies' ch. 11 of G. Blazyca and R. Rapacki eds, *Poland into the New Millennium*, Edward Elgar, Cheltenham.

GUS, (1998), *Polska w nowym podziale terytorialnym. GUS.* Warszawa.

Klasik, A, (2001), *'Polityka regionalna państwa i polityka rozwoju województw - refleksja strategiczna'*, (Regional policy of the state and voivodships - strategic reflection) in *Polityka regionalna państwa po!śród uwikłań instytucjonalno-regulacyjnych*, J. Szomburg (ed), Instytut Badań nad Gospodarką Rynkową, Gdańsk.

Kołodziejczyk, D, Wasilewski, A, and Lidke, D, (1998) *Rozwój demograficzno-gospodarczy w skali lokalnej*, (Demographic and Economic Development on the Local Scale), Instytut Ekonomiki Rolnictwa i Gospodarki Żywnościowej, Warszawa 1998.

Kozak M., and Pyszkowski A., (2000), 'Uwarunkowania rozwoju regionalnego Polski I postawowe dylematy polskiej polityki regionalnej' in *Polityka regionalna i jej rola w ponoszeniu konkurencyjności regionów*, Klamut M., and Cybulski L., eds Wydawnictwo Akademii Ekonomicznej we Wrocławiu, Wrocław.

Kudłacz, T, (1999), 'Zróżnicowanie rozwoju regionalnego w Polsce na tle sytuacji w Unii Europejskiej' (Regional differentiation in Poland in EU context) in *Polska w Unii Europejskiej* ((Poland in the EU), Instytut Studiów Strategicznych, vol. 39, Kraków 1999.

Kudłacz, T, (2001), 'Rozwój regionalny Polski lat 90. - ocena dominujących procesów oraz spodziewanych tendencji' (Regional development in Poland in the 1990s - evaluating major processes and expected tendencies) in *Polityka regionalna państwa po!śród uwikłań instytucjonalno-regulacyjnych*, J. Szomburg ed., Instytut Badań nad Gospodarką Rynkową, Gdańsk.

Orłowski,W, Saganowska, E, Żienkowski, L, (1998), *Szacunek produktu krajowego brutto według województw za 1996 i 1997 rok.* (Estimates of GDP by voivodship in 1996 and 1997), ZBSE GUS i PAN Warszawa.

Orłowski, W.M, (2001), 'Polskie regiony na tle wyzwań integracyjnych' (Polish regions in the context of integration challenges), in *Polityka regionalna państwa pośród uwikłań instytucjonalno-regulacyjnych*, Szomburg J., ed, Instytut Badań nad Gospodarką Rynkową, Gdańsk.

Pisz, Z, (2001), *Problemy społeczne transformacji*, (Social problems of transformation)

Akademia Ekonomiczna im. Oskara Langego we Wrocławiu, Opole.

Rzeczpospolita, (1998), 'Informacja o dochodach województw i powiatów. Ministerstwo Finansów' May 26,

Surażska, W, (1999), 'Gdzie Polska rozwija się najszybciej', (Where Poland is developing fastest), *Rzeczpospolita*, September 17.

Winiarski, B, (2000), 'Zróżnicowania poziomu konkurencyjności regionów a kierunki i cele polityki regionalnej oraz jej uwarunkowania makroekonomiczne w Polsce' in *Polityka regionalna i jej rola w podnoszeniu konkurencyjności regionów,*M. Klamut and L. Cybulski (eds), Wyd. Akademii Ekonomicznej we Wrocławiu, Wrocław.

Żienkowski, L, (1997), 'Why do regional gross products differ?', *Research Bulletin RECESS (ZBSE)*, GUS, Warsaw, no. 4.

Chapter 5

The Economic Restructuring of the West of Scotland 1945-2000: Some Lessons from a Historical Perspective

John Foster

Introduction

The thesis of this paper is simple. In 1945 the West of Scotland comprised an industrial district. It possessed significant competitive advantages in the production of transport and power equipment: turbines, pumps, aircraft engines, generating equipment, mining equipment, guidance systems, rail locomotives and ships. Restructuring destroyed it – and has failed to replace it.

This said, this paper has no intention of being simplistic. The industrial district as it existed in 1945 was in incipient crisis. It required reorganization. The problem was that the three successive phases of restructuring, in the late 1940s, in the late 1950s, and again from the late 1970s, were driven by agendas that were external. The needs they addressed were not those of Clydeside but of British business at UK level.

Arguing this case is quite the reverse of being simple. It requires a narrative that covers economic history, industrial and regional policy for four decades, a reconstruction of the internal politics of business at British and regional level and some understanding of capital's changing relationships with labour. Our treatment will therefore have to be schematic. We will start with a comparison of the region's industrial and occupational structure in the 1940s and 1990s and consider its relative strengths and weaknesses. This will be followed by a schematic assessment of the three stages of restructuring. We will conclude with a consideration of alternatives and the possible lessons.

Then and Now: The Destruction of an Industrial District

Alfred Marshall developed his concept of the industrial district in the 1890s when Clydeside was at the height of its industrial expansion. Marshall sought to explain the continuing entrepreneurial vigour of Britain's regional economy in an age of growing concentration. His explanation was in terms of the potential for significant external economies of scale and scope when large numbers of similar firms were located in the same district. Knowledge was exchanged. Specialist suppliers shared skills that were primed by innovation in the core firms. Skilled labour disseminated good practice.

Over the past decade the rise of institutional economics, and the new policy significance of the region, have refocused attention on the ability of some locations, as against others, to sustain innovative, research-led growth in a particular speciality. 'The learning region' has become a central concern of policy makers. Research on the characteristics of such industrial districts as they exist in the later twentieth century is now beginning to identify a number of factors. They are:

- the level of specialization in linked industries: the higher the better (Krugman 1991; Dalum 1999);
- the volume of specialist labour (rather than artificially-created networks for the exchange of knowledge) (Simmie 1997);
- the existence of regionally-based core firms, with significant R&D, and the active character of their productive links with smaller firms (rather than any general homogeneity of business culture) (Mueller and Loveridge 1999; Lazarson and Lorenzoni 1999; Sternberg and Tamasy, 1999);
- the presence of institutionally-embedded and path dependent learning routines that evolve over long time periods and cannot quickly be imitated elsewhere (Maskell and Malmberg 1999);
- the existence of a relative geographical stability for core competences which encourages long-term investment in training rather than reliance on short-term contracts and labour flexibility (Michie and Sheehan, 1999, Nooteboom, 1999, Anderson and Tushman, 2001).

With some qualifications it is argued that this is what Clydeside possessed in 1945 and what it has lost since.

In the Clyde Valley Regional Plan of 1946 Sir Patrick Abercrombie stressed the importance of building on the region's existing specialisms.

> Shipbuilding is essentially an assembly industry where an enormous number of parts of the highest quality are needed, varying from steel plates, propellers and steering gear to wireless equipment, furniture, cooking appliances, refrigerators and paint. ... There has therefore grown up with the shipbuilding industry a vast assortment of firms (mostly in the Glasgow area) catering primarily for the demands of the shipyards and catering primarily for its engineering and furnishing requirements ....

On the basis of this concentration Clydeside shipbuilding still possessed significant advantages. In 1938 Clydeside produced 40 per cent of UK and 15 per cent of world shipping output. Between 1945 and 1950 its percentage of world output increased to 20 per cent – and not just because of the temporary dislocation of other suppliers. Shipbuilding in turn provided the major consumer for steel – with Clydeside producing 2.29m tonnes in 1948 (over 12 per cent of the UK total). The demand for turbines and ships' engines also interlocked with other local specialisms in generating equipment, aero-engines and pumps. In all these areas Clydeside remained highly competitive at a world level (Abercrombie 1946/1949).

As a planner working in the tradition of Patrick Geddes, Abercrombie saw the importance of maintaining evolutionary development and combining existing strengths with some admixture of newer, mass production industry. He also believed that this process needed to be planned on a regional basis – and that for optimum results a measure of direct state intervention in production would be required. State ownership was already the case with coal and aircraft production. He saw it highly desirable in steel – where only government planning could secure the industrial reorganization required for the giant tidewater plant at Erskine proposed in the Brassert report (Brassert, 1929). Abercrombie also noted problems with the existing patterns of ownership and control in shipbuilding. The stress on quality had tended to militate against the introduction of mass fabrication techniques. The plethora of family-owned firms, still over twenty in number on the Clyde, was unlikely to be able to respond to the opportunities that would be opened up by a new steel complex at Erskine. Reorganization of the industry on the basis of a few big yards lower down the Clyde would probably require state ownership.

Table 5.1   Industrial Employment in Clyde Valley and Strathclyde in 1939 and 1990
('000 employees)

Industrial Group	**1939**	**1990**
Mining	47	0
Metal manufacturer	35	6
Other metal	26	7
Engineering	70	40
Shipbuilding	57	15
Instrument engineering	0	4
Computers	0	6
Textiles and clothing	80	24
Leather	3	1
Furniture	13	4
Brick, pottery, glass	9	5
Chemicals, plastic	15	11
Food	40	28
Other manufacturing	9	2
Total	404	153

Source: C. H. Lee, Scotland and the UK, 1995, p. 61

What happened? Thirty years later, in 1985, Peter Payne wrote the obituary of heavy industry on the Clyde. His conclusions highlighted what had not been done. The failures were precisely the challenges identified by Abercrombie forty years before. They were the failure to innovate in modern steel production capacity in the 1940s, the lack of investment in new shipbuilding techniques during the postwar boom years and the failure to utilise existing strengths in the development of new products (Payne 1985). Table 5.1 shows the shift in employment patterns between 1939 and 1990 for approximately the same area of the Clyde Valley. By the 1990s Clydeside had more or less totally lost its earlier specialization in engineering. As a region it is now more dependent on services than the rest of the UK – and services without a secure and vigorous production base tend to be technologically limited (Capron, 1997). Its one area of specialism is in the assembly of office equipment and IT. This is almost entirely externally-owned and controlled. It might conceivably provide the future basis for an industrial district with a world competitive edge. But it certainly does not at the moment and it appears increasingly unlikely that it ever will (Turok 1997; McCann, 1997; McGregor 2001). Writing in 2000 David Webster warned of the deepening crisis facing the deindustrialized core of Clyde Valley. Its population, he said, was falling fast - and its potential for economic recovery fading with it. Long term unemployment had robbed the area of the young and the skilled, and the projected decline in the size and earning capacity of the working population was sufficiently serious to threaten the future fiscal basis for regional services and infrastructure (Webster 2000). We will return to these points at the end of the paper.

Table 5.2   Employment in Scotland in 1951 and 1993
        (percentage of total)

Employment Category	1951	1993
Agriculture	7.3	1.3
Energy, water	5.7	2.6
Metals, chemicals	5.6	1.8
Metal goods, engineering	12.6	7.7
Other manufacturing	16.9	8.7
Construction	6.8	5.4
Distribution	12.4	20.7
Transport, communications	8.2	5.5
Finance	1.5	10.2
Public services	24.2	35.1
Total	100.0	100.0
Total ('000)	2,195	1,984

Source: C. H. Lee, *Scotland and the UK*, 1995 p. 66.

Keeping the Politics in Political Economy

Had Abercrombie's plans been put into operation there is at least a chance that Clydeside would still possess, as a region, some of the competitive advantages enjoyed by producers of capital goods elsewhere in Europe and America. However, Abercrombie's type of planning never had a major constituency of support in government. His patron, Tom Johnston, who commissioned the Clyde Valley Plan, retired as Secretary of State for Scotland in 1945. At Cabinet level the same period saw the advocates of Keynesian demand management comprehensively defeat perspectives for a much more directive form of regional economic planning as argued by Dalton and Cripps. The problem with the resulting restructuring was that there was no overriding regional perspective. It was instead a series of ad hoc responses to externally conceived agendas largely set by the business and banking establishment at a UK level – reflecting needs which were often in direct conflict with the coherent development of the regional economy.

Why was this regional perspective absent? It was not the result of any fundamental change in the relationship between business and the British state. Banking and industry had played a central role in British policy formation over the previous century. What changed was the scale of economic concentration and the type of linkages that existed between industry and the banking system. In the 1920s and 1930s regional concentrations of industry and regional banks enjoyed considerable autonomy across Britain. In Scotland the leaders of heavy industry and the associated Scottish banks had enjoyed a close alliance with the Bank of England and the government. What happened during and immediately after World War II was a sea change in these relationships. This was so in terms of the scale of industrial activity and in the articulation of relations with the City of London banks. More especially it was in the character of external, international, economic alliances. The maintenance of sterling as a world banking currency and the adoption of full employment Keynesian economics demanded an alliance with the United States. This in turn meant promoting the growth of multinational firms that were sufficiently large and competitive to survive within the US trading area. These firms operated at UK level. They were mainly financed and controlled from the City of London and were primarily in motors, petrochemicals and electrical engineering.

For the following thirty years both Labour Party and Conservative governments remained wedded to this perspective. The key requirement placed on the UK growth industries was to maximise exports in order to feed dollars back into the economy and to rebuild the pound sterling as a world banking currency. The alliance with the United States in turn placed significant constraints on the freedom of action of post-war governments. Some of these constraints were particularly detrimental to Clydeside. Traditional markets for Clydeside produce in the Sterling Area and Commonwealth were opened to American competition. Marshall Aid tied assistance to the use of American ships. Especially after the sterling crisis of 1947 steel quotas were directed towards the new 'export' industries in the south-east. In so far as the nationalization of transport and power provided the opportunity for dramatically lowering infrastructure costs, this infrastructure tended to be geared towards the needs of the new generation of UK multinational companies (Foster 1986 and 1993; Harvie 1983).

This wider background is important. Agendas for restructuring are not neutral. They serve interests. Of course the programmes themselves, for reasons of credibility, will always be argued with an apparently neutral economic rationale. They will additionally often have quite unintended consequences. But unless the wider politics are understood it is not possible to appreciate the intent itself. The three phases of restructuring on Clydeside correspond to increasing levels of economic concentration at British level. Each was matched by an *increasing* level of state intervention. The third stage, initiated in the 1980s and largely continuing today, extended to detailed attempts at social re-engineering and restructuring at micro-economic level. Some aspects of this third stage are explored elsewhere in this volume in the paper presented by Chik Collins (see ch. 6) and reveal the remarkable concern manifested by governments to control the local labour infrastructure and its organization (Collins 1999). Each of the three stages matches changes in external agendas. But they also demonstrate, as we will see, an internal dynamic. The unintended consequences increasingly modified and, in the third stage determined, the character of the subsequent restructuring.

Three Phases of Restructuring

The Creation of a Dual Economy: 1947 to the Late 1950s

The period was marked by the failure of government to implement the Clyde Valley Plan, a policy of malign neglect as applied to the shipbuilding industry (both by the government and its owners) and the use of the West of Scotland as a parking ground for American branch plants. The result was the creation of a dual economy.

The American government's post-1947 policy of economic and political stabilization in Europe depended in large part on direct investment by American multinationals. Britain was the most favoured location. But British governments did not particularly want these new factories – at least not in the south east. They were fearful of the inflationary pressures that would result if US firms, seeking skilled labour and generally paying higher wage rates, were located alongside British producers of cars and consumer durables. US investors were therefore directed overwhelmingly to Scotland. By 1952 Scotland had more US investment than the rest of Europe put together and the highest per capita US investment outside Canada.

Abercrombie had called for the creation of New Towns. He saw them as a way of decongesting urban areas. They would develop organically as industry was shifted out of the cities. In the event the New Towns became the geographical core of this dual economy: the favoured environment for American investment. American investors wanted union-free sites that were well away from the old industries. Clydeside industrialists also wanted the US branch plants to be at as safe a distance as possible.

Back in the old industrial areas the owners did not reinvest their high post-war profits. They had no incentive to do so. Raw material quotas and export credits went elsewhere. Trade and aid agreements between the British and American governments meant that UK shipbuilding would have a limited market once the post-war building boom ended.

So instead of comprehensive redevelopment, there were two largely separate economies. No tidewater steel plant was built and there was no reorganization of shipbuilding – at just the time when this was taking place in competitor regions (Robertson 1996).

The 1960s: Dual Economies at War

The late 1950s and early 1960s saw the beginning of a far more comprehensive intervention by central government in the regions - beginning with Scotland. This new policy had its origins in the sterling crisis of 1957-1958. Inflationary overheating in the south-east had resulted in a sharp rise in interest rates. Government advisers argued that the German and Japanese economies were able to secure steady, inflation-free growth because of their access to rural or migrant labour. While this type of surplus labour was not available for Britain, it was proposed that the same effect could be secured by industrial relocation. Production could be progressively shifted into the regions. In Scotland, the North East and the North West unemployment levels were far higher than in the south-east.

This was the origin of what Doreen Massey has called 'surplus labour regional policy'. Regional development now became central to British macro-economic policy. Again, however, the beneficiaries were to be the big British multi-nationals. By siting their new capacity in the regions it was believed that these firms would be able to grow at the same rate as their rivals overseas. A new language emerged. Pre-existing industries were stigmatized as 'declining'. The new branch plants represented 'modernization' (Massey 1984).

The Conservative government of Harold Macmillan set to work to create the detailed conditions for this radical restructuring. In Scotland the first step was to move up a major British car producer, Rootes, and to create local supplies of strip steel. This brought a major conflict with Colvilles, the one remaining Scottish steel producer, mainly controlled by shipbuilding interests and mainly geared to producing shipbuilding plate. Colville's board knew that an expansion into strip steel would place severe pressure on supplies of coking coal, steel scrap and labour and increase the unit cost of plate. They refused. A period of fairly intense conflict resulted. Macmillan changed the financial regulations for shipping credit so that British ship owners only received credits if they bought their ships outside the UK. He threatened to finance an English steel firm to move into Scotland and establish a strip steel plant at Grangemouth. Eventually a compromise agreement was secured by which Colvilles built a small steel strip mill at Ravenscraig (Payne 1985; Foster 1986).

The key feature of this period is the explicitness with which policy-makers justified a fairly root and branch destruction of the old infrastructure. The government-sponsored Toothill Report of 1961 was chaired by the Scottish managing director of Ferranti and had no representatives of Scottish engineering. The report advocated a new industrial culture based on mass production, not batch production, developing from growth points that would be created around the New Towns. The assumption was that growth would depend on new incoming firms from the south. At British level the new National Economic Development Council (NEDC) set out targets for the transfer of labour from 'old' industries to the new. At a British level the 1962 NEDC report estimated that 200,000 workers could be transferred annually to fuel growth in the new industries. Declining industries were listed as coal, steel, shipbuilding and railways. Because these industries were either state controlled or quite closely dependent on the state, the government was able to ensure that this happened. Between 1960 and 1967 some 119 of the 166 coal-pits in Scotland were closed and 30,000 mining jobs lost. Closures of a similar scale happened on the railways. In shipbuilding 23,000 jobs went over the same few years and five major Clydeside yards were closed.

How many new jobs were created? Something like 24,000. Few of the incoming plants survived for more than twenty years. Most were what George Kerevan described as 'cathedrals in the desert'. They imported supplies. They had little contact with the surrounding economy. They were dependent on core production elsewhere – either in the south-east or the US. As the world economy went into recession from the 1970s, they were among the first to go (Woolfson and Foster, 1988).

There was also a more fundamental problem, an unintended consequence which largely determined the character of the third phase. Government assumptions about the availability of surplus labour in the regions proved to be seriously flawed. While unemployment percentages were high, the absolute numbers of skilled workers in the right place at the right time were quite small. Inflationary pressures began to be felt on Clydeside from the early 1960s. American firms in particular tended to bid up wages. By 1970 the wage differential between Clydeside and the south-east had been lost. This placed further pressures on the remaining regional engineering firms. It also created the context for the redevelopment of a highly militant trade union movement. When the Conservative government moved to close the remaining shipyards on the upper Clyde in 1971, the 8,000 workers occupied the yards. In doing so they succeeded in creating a much broader social alliance. This alliance encompassed most of Clydeside's small and medium enterprises, Scottish local authorities and the wider trade union movement. Its unifying demand was a comprehensive reversal of economic policy to save what remained of Clydeside as an industrial district. In the event the Conservative government reprieved the shipyards and offered substantial aid to the regional economy.

This period was a critical one for the future. In the mid-1970s a full reversal of policy seemed possible. The type of regionally-accountable democratic planning envisaged by Abercrombie was once more politically in the frame. In the aftermath of the shipyard occupations the *Scottish Trades Union Congress* convened Scottish Assemblies in 1972 and again in 1973. These brought representations from all sections of Scottish life and endorsed perspectives for economic redevelopment under the democratic control of a Scottish Parliament. The Labour Party in turn adopted this policy. When returned to power in 1974 the new Labour government established a *Scottish Development Agency* and took into public ownership shipbuilding and aircraft construction, nationalised the aero-engine producer Rolls Royce in time to rescue the RB 211 engine, and started to establish a public sector holding in the offshore oil industry. In 1978 Labour made a first bungled attempt to establish a Scottish parliament.

Ultimately, however, events took a quite different turn. The Conservative Party retained in England the political base it had already largely lost in Scotland. Out of office after 1974 it fundamentally reshaped its policies away from corporate Keynesianism to neo-liberal monetarism. A major part of its post-mortem concerned issues of ideology and the control of labour - focused especially its experiences in Scotland (Foster 2001).

1979 On: From Shock Treatment to Micro-Economic Intervention

It was this that provided the context for the third phase of restructuring. The Conservatives returned to power in 1979 with a radical agenda to solve once and for all the problems of labour militancy and wage inflation in British industry. They planned to use the windfall of North Sea Oil to effect a massive market clearing operation. As the pound became a petro-currency, it would be allowed to rise in international value. Capital would be exported on preferential terms. Very large-scale unemployment would end the 'cycle of rising expectations', permit the historic defeat of the trade union movement and then allow the repatriation of capital on its own terms.

These plans were worked out in some detail before the election. They targeted in particular those industries with large-scale corporate style management where union power was strongest. They also included some measures of social engineering such as the sale of social housing – believed to provide the social basis for collectivist attitudes. But the main reliance of the scheme's architects, principally Keith Joseph and Nicholas Ridley, was on the brute force of mass unemployment across the whole country (Woolfson, Foster and Beck 1997).

Unemployment was achieved. But it proved much more difficult to defeat the trade union movement – and the cost of doing so was an unexpectedly high level of political radicalization (especially in the inner-cities and the regions). It also proved very difficult to get capital back into Britain. From the early 1980s a subsidiary agenda began to emerge. This involved much more detailed micro-economic intervention. 'Japanization' became the model (the first mention of the 'sunrise' industries in the *Economist* was in 1981). This meant the encouragement of new workplace relations and new core-periphery relations between large and small firms. In Scotland, where the Conservatives had suffered heavy losses in both 1982 and 1987 general elections, the government also initiated ambitious experiments in the social re-engineering of urban working class communities. The emphasis was now on appropriate supply-side factors and in particular flexible labour.

As government advisers studied the mechanics of Japanese industry and sought to attract external investors, the pendulum swung back towards external economies and the recreation of industrial districts. The difference was that Scotland now had virtually no core firms of its own around which industrial districts could naturally develop. The entire process had to be simulated. It required a scale of detailed government intervention that would have been unthinkable even thirty years before. Scottish level agencies levered in new core companies from the US and Japan to establish branch production facilities. Local Enterprise Companies (the so-called LECs) sought to restructure inter-firm relationships between local small firms and new incoming firms. 'Local Economic Initiatives' were set the task of preparing appropriate labour. Highly ambitious projects were launched to transform the mentalities of the working population. The most bizarre was the introduction of a poll tax system of local government taxation. This was intended by its inventors to individualise voting in council elections on a personal cost-benefit basis and thereby destroy the class solidarity basis of Labour support. As an attempt to interest Scots in the game theory aspects of economic rationality, it did not meet much success. After three years it was abandoned amid a massive campaign of civil disobedience (McConnell 1994). Another more long-lasting experiment in social engineering was *New Life for Urban Scotland* adopted at the same time and described by Chik Collins in chapter 6 of this book. This was focused initially on the most recalcitrant of working class neighbourhoods - often containing potentially important reservoirs of the type of flexible female labour required for incoming employers. These 'Social Partnership' projects sought organizationally to penetrate local neighbourhoods and substitute ultimately state controlled networks for the existing autonomous tenants and community organizations. Again the results were sometimes bizarre and unexpected – although in essence the same types of project continue in a new and expanded form today.

Consequences

We have already reviewed some of the consequences of the three phases of restructuring identified above. The production sited in Scotland by overseas electronics and IT firms was predominantly low skill assembly and the resulting labour profile was quite distinct from that in the south and east of England (McNicoll, 2000). When recession struck in 2000-2001, these plants were the first to close – sometimes with work being transferred to Eastern Europe where labour costs were lower still. Even at their best the electronics and IT industries only provided a fraction of the required jobs. In the de-industrialised core of the West of Scotland the situation was considerably worse. The demographic crisis was only one aspect of the problem. Its severity was framed within a wider disintegration. Young people in these areas have some of lowest levels of educational achievement in Britain. The incidence of illness and morbidity is the worst in Western Europe. Crime, drug abuse and anti-social behaviour now make parts of Glasgow and surrounding conurbations almost hostile environments for business growth.

The intent behind the micro-economic interventions of the 1980s and 1990s was of course quite the opposite. The objective was to create a compliant labour force and a local pro-business culture of partnership and individualism. One part of this objective was indeed met: long-term unemployment and the closure of the old unionised workplaces has resulted in a severe weakening of any form of class organization. Only 22 per cent of workers in Scottish manufacturing are today trade union members – the lowest percentage anywhere in Britain. The same disintegration has also affected neighbourhood organizations. Yet the vacuum has not been filled with the promised entrepreneurial values. More seriously, the old industrial culture of pride in skill and craft has been severely eroded for a very significant section of the population. Michie and Sheehan have recently demonstrated an association between innovation, productivity growth and the existence of a stable, confident and (generally) unionised workforce. While the skills of the 1950s are certainly not those required for the 2010s, no effective context was created for their development and renewal.

In analysing the sharp differences of achievement among transition economies, Vladimir Popov has recently pointed to the positive importance of institutional factors – particularly continuing state expenditures on infrastructures, the maintenance of elements of labour discipline and the firmness of macro-economic state control. Where these are present, growth has been positive. Where absent, transition has led to severe economic and social regression (Popov 2000 and 2001). While it would be fanciful to describe Clydeside as a transition economy, its experience of restructuring since the 1950s seems to provide some interesting and worrying parallels.

Some Lessons

Was there an alternative? The argument here is that there was. The proposals put forward by Abercrombie offered a good start. Initially, in 1946, attempts were made by the *Scottish Office* and the *Board of Trade* to implement plans for the structured development of the region's heavy industry. The most notable was the bid to take control of the heavy pressings plant at Linwood from the Lithgow-Colville group (Sims 1984).

Given the resistance by the owners, some level of public ownership would probably have been necessary (Michie and Prendergast 1998). There seems no reason to believe that public ownership would not have been successful in this field. The record of Britain's post-war nationalised industries provides no evidence that it was less efficient than private industry (Millward 1997; Millward and Singleton 1995). There is some evidence that it would have handled innovation effectively – and certainly more effectively than the Clydeside owners (Sawyer and O'Donnell, 1999).

The irony is that by 1971 Scotland had become the site for the biggest engineering project anywhere in the world. Offshore oil took one fifth of all Britain's industrial investment for a decade, and almost all of it was in engineering construction work. In the event only a minority of this work was undertaken by Scottish firms and virtually none of the steel was produced in Scotland (Woolfson, Foster and Beck 1997).

In summing up what was wrong with Clydeside's restructuring, David Webster provides a starting point:

> It is argued here that the distress of Glasgow and the Inner Clyde Valley is to a substantial extent actually the result of policies dating from the 1960s. Scottish policy makers adopted a particularly strong version of the 'growth pole' strategy by concentrating resources on the New Towns. ... [the] consequence has been to cause economic decline in the unfavoured areas without sufficient complementary growth in the target areas. (Webster 2000).

Some of the difficulties might be listed as:

- reliance on ultimately erratic and unpredictable market forces – with incoming firms often neither the most suitable in terms of their production objectives nor the most robust in terms of world position;

- fragmented implantation – with incoming firms dictating their terms for infrastructure;
- displacement of an existing industrial culture – both in terms of micro-interventions to change the alignment of SME towards external firms and the fairly deliberate dispersal of a pre-existing workforce with very high level engineering skills.

In today's world where economic restructuring is increasingly prescribed at an international level, there would, therefore, seem to be significant lessons to be learnt. Of these the most important is the need for some democratic platform from which to interrogate external agendas and particularly to ensure that they complement and do not destroy what is of economic value within the regional infrastructure.

References

Abercrombie, P. (1949), *The Clyde Valley Regional Plan 1946*, HMSO, Edinburgh.

Anderson, P. and Tushman M. (2001, 'Organisational environments and industry exit', *Industrial and Corporate Change*, Vol. 10/3, September.

Brassert, H. *Report to Lord Weir of Eastwood on the Manufacture of Iron and Steel*, May 16, 1929, (British Archive, Scottish Regional Records Centre, Glasgow).

Capron, H. (1997), 'Role of the manufacturing base in the development of services', *Regional Studies*.

Collins, C. (1999), *Language, Ideology and Consciousness*, Ashgate, Andover.

Dalum, B. et al (1999), 'Does specialisation matter for growth', *Industrial and Corporate Change*, Vol. 8/2.

Foster, J. and Woolfson, C. (1986), *The politics of the UCS work-in: class alliances and the right to work*, Lawrence and Wishart, London.

Foster, J. 1993, 'Labour, Keynesianism and the Welfare State', in J. Fyrth (ed.), *Labour's High Noon: the government and the economy 1945-1951*, Lawrence and Wishart.

Foster, J. (2001), 'The twentieth century', *New Penguin History of Scotland*, W. Knox and R. Houston (eds.), Penguin.

Harvie, C. (1983), 'Scottish Labour and world war II', *Historical Journal*, Vol. 32.

Krugman, P. (1991), *Geography and Trade*, MIT.

Lazerson, M. and Lorenzoni, G. (1999), 'The firms that feed industrial districts', *Industrial and Corporate Change*, June.

Lee, C.H. (1995), *Scotland the United Kingdom*, Manchester University Press, Manchester.

Maskell, P. and Malmberg, A. (1999), 'Localised learning and industrial competitiveness', *Cambridge Journal of Economics*, Vol. 23/2, March.

Massey, D. (1984), *Spatial Divisions of Labour*, Macmillan, London.

McCann, P. (1997), 'How deeply embedded is Silicon Glen? A cautionary note', *Regional Studies*, Vol. 31/7.

McConnell, A. (1994) *State Policy Formation and the Origins of the Poll Tax*, Dartmouth, Aldershot.

McGregor, P. 'FDI in the electronics sector', *Quarterly Economic Commentary*, January, Vol.25/1.

McNicoll, I. (2000), 'Industrial demand for skilled labour: Scotland and the UK', *Quarterly Economic Commentary*, XXV/3.

Michie, J. and Prendergast, R. (1998), 'Government intervention in a dynamic economy, *New Political Economy*, Vol. 3/3.

Michie, J. and Sheehan, M. (1999), 'HRM practices, R&D expenditure and innovative investment: evidence from the UK WIRS', *Industrial and Corporate Change*, Vol. 8/2 June.

Millward, R. and Singleton J. (1995), *The Political Economy of Nationalisation in Britain, 1920-1950*, Cambridge University Press.

Millward, R. (1997), 'The 1940s nationalisations: means to an end or means of production', *Economic History Review*, Vol. 50/3.

Mueller, F. and Loveridge, R. (1995), '"The second industrial divide?" The role of large firms in Baden-Wurttenberg', *Industrial and Corporate Change*, Vol. 4.

Nooteboom, B. (1999), 'Innovation, learning and industrial organisation', *Cambridge Journal of Economics*, 23/2.

Payne, P. (1985) 'The decline of Scottish heavy industries', in R. Saville (ed.), *The economic development of modern Scotland*, John Donald, Edinburgh.

Popov, V. 'Lessons form the transition economies: strong institution are more important than the speed of reforms', UNRISD conference September 2001 based on Popov in *Comparative Economic Studies*, 2000, vol. 42/1.

Robertson, D. (1996), 'Scotland's New Towns', in A McInnes (ed.), *Scottish Power Centres*, Edinburgh.

Sawyer, M. and O'Donnell, K. (1999), *A future for public ownership*, Lawrence and Wishart, London.

Simmie, J. (1997), *Innovation, networks and learning regions*, Kingsley, London.

Sims, D. (1984), *Car manufacturing at Linwood: the regional policy issues*, Departmental Paper (Politics and Sociology), Paisley College/University.

Sternberg, R. and Tamasy, C. (1999), 'Munich as German's No.1 high technology region', *Regional Studies*, Vol. 33/4.

Toothill, J. (1961), *Inquiry into the Scottish economy*, Paton, Paisley.

Turok, I. (1997), 'Linkages in the Scottish Electronics Industry', *Regional Studies*, Vol. 31/7.

Webster, D. (2000), 'The Political Economy of Scotland's Population Decline', *Quarterly Economic Commentary*, Vol. 25/2.

Woolfson, C. and Foster, J. (1988), *Track record: the story of the Caterpillar occupation*, Verso, London.

Woolfson, C. Foster, J. and Beck, M. (1997), *Paying for the Piper: capital and labour in Britain's offshore oil industry*, Mansell, London.

PART III

SOCIAL PARTNERSHIP AND REGIONAL ECONOMIC DEVELOPMENT

Chapter 6

Urban Policy, 'Modesty' and 'Misunderstanding': On the Mythology of 'Partnership' in Urban Scotland

Chik Collins[1]

Introduction – Urban Policy, 'Modesty' and 'Misunderstanding'

Some time around the late 1980s it became increasingly common for commentators to point to a framework of urban policy in Scotland which had distinctive characteristics in comparison to the rest of Britain (see for example Atkinson and Moon, 1994; Boyle, 1988; McCarthy, 1999; McCrone, 1991). But if there have been distinctively Scottish characteristics in this policy field over the past fifteen years, then it would seem that modesty has not been among them. Rigorous evaluation of, and sober reflection upon, the experience of implementation would, unfortunately, not be among them either. For what was perhaps all too characteristic of urban policy in Scotland during that period was a tendency towards exaggerated claims about its efficacy, together with an unwillingness to face the reality of what comes close to (or maybe just *is*) policy failure. For those who might want to learn from the experience of Scotland in recent times, especially those who might be looking for some transferable policy 'know-how', this might create some difficulties – and some dangers.

A couple of examples come immediately to mind. In the late 1980s and during 1990, Ferguslie Park, a very deprived area of Paisley, was a participant in a European Community project called *Areas in Crisis*. The areas focused upon were those where 'the question of the disintegration of the urban fabric' was posed. At that time Paisley's Ferguslie Park had recently been designated as an urban policy 'Partnership Area' under a new government programme labelled *New Life for Urban Scotland* (Scottish Office, 1988). The aim of the Ferguslie Park Partnership (FPP) was to transform the area, within ten years, from one that was 'disintegrating' to one that

[1] I am grateful to John Foster and Jim Lister for comments on an earlier draft of this chapter. I have benefited too from discussions on 'partnership' with participants in the Big Issue Foundation Scotland Conference on 'Social Inclusion Partnerships: How Should Local Communities Respond?' (31 May 2000), (Velasco, 2000). Parts of the paper have been published as 'From New Life to New Labour: The Mythology of Partnership' in Concept (The Journal of Contemporary Community Education Practice Theory), 11, 2, 2001, pp. 3-7. The views expressed here are however those of the author alone.

would be seen as 'a desirable place to live'. Within the *Areas in Crisis* project, the view put across by the participants from FPP was that in Scotland policy makers had learned the lessons of past failures and had created a model capable of solving deep-seated urban problems. This was a model that others in Europe should aspire to emulate. This view was largely taken on board by the authors of the *Areas in Crisis* report (Jacquier, 1990a, see also Jacquier 1990b). Yet, within five years of these reports being published, the name Ferguslie Park had been projected nation-wide (i.e. across the whole UK) as being synonymous with all that was worst about the problems of 'social disintegration' in poor working class areas – drug selling, money lending, gang violence and murder. To make matters worse, much of the above had been perpetrated, it was alleged, through the offices of a so-called 'community business' established through the intervention of the FPP itself. Analysts have been slow to trace the linkages between the operation of the FPP more generally and this particular policy failure, but a little local knowledge (which is often what urban policy researchers lack) means that they can be traced quite clearly (Collins, 2000).

Some commentators have been keen to dismiss the Ferguslie Park experience as a 'one off'. But, in the sense that the Scottish policy community has tended to make rather immodest claims about its own efficacy, it may be rather typical. A very good example is given by María Gómez (1998) in her analysis of the relationship between urban regeneration policy in Glasgow and Bilbao. The re-marketing of Glasgow in the 1980s and 1990s involved emphasizing how much the city had changed for the better. This story of Glasgow's great 'success' – its transition from industrial decay to an attractive and vibrant location for the service sector – was, however, simultaneously a story about the great 'success' of policies (and policy-makers). In this way Glasgow's policies were also portrayed as a model for other de-industrializing cities to follow. In the city of Bilbao the policy-makers sought to do just that. The problem, for Gómez, is that the story of Glasgow's apparent success is a misleading one, and that it inspired, and was used to legitimize, urban renewal policies in Bilbao that were not likely to bring any more real 'success' in that city than in Glasgow. Comparing the approach in Bilbao with that in Glasgow she observed:

> It is not only the same idea, but the same message and even the same words. The ... Metropolitan Bilbao Plan ... repeatedly insists on the need to follow the path of Glasgow. (1998, p. 113).

She concludes that in Bilbao the 'utilization of Glasgow's misleading regeneration clichés' betrays, at best, 'a clear misunderstanding of Glasgow's process by Bilbao's decision-makers' (1998, p. 118). However there can be little doubt that policy-makers and agencies in Glasgow, and in Scotland more generally, have contributed to this 'misunderstanding', and in doing so have perhaps also provided legitimation for some who might understand perfectly well the nature of Glasgow's process.

The lesson that seems to emerge from this is that those who would seek to learn from the experience of urban policy in Scotland would do well to look sceptically on the claims that have been made, and are still being made, about its efficacy. This chapter provides a short case study of such scepticism. It presents a critical appraisal of the most prominent urban policy initiative in Scotland in recent times – the *New*

Life for Urban Scotland programme launched by the Conservative Government in the wake of Mrs Thatcher's third general election victory in 1987. This was the programme that established the model of 'partnership' that has been at the heart of virtually all subsequent action in the field. The central elements of this model were endorsed by the New Labour administration after the 1997 General Election, and have provided the basis for the Scottish Executive's[2] current policy in the field – its *Social Inclusion Partnerships*. In this sense the transition from *New Life* to New Labour has been almost seamless.

Visitors to Scotland might be impressed by the strength of the consensus surrounding the supposed efficacy of this 'partnership' model. It might appear that we Scots have a model that has been tried and tested, and found to be effective. A study of press coverage on the subject might heighten the feeling of confidence. Whereas the media is somewhat sceptical regarding claims about the value of policies in other areas, in urban policy the coverage adopts, fairly uncritically, the narratives of the policy makers and implementers – often tales of heroic efforts and success against the odds. However, on closer scrutiny this consensus seems to rest on some rather dubious foundations. There seems to have been yet more of the kind of 'misunderstanding' identified by Gómez – though on this occasion within Scotland itself. And in order to make this case we need look no further than the official evaluation of the *New Life* 'partnerships' (Cambridge Policy Consultants, 1999). We will, though, if we are to avoid further 'misunderstandings', have to look beyond the 'spin' it seeks to put on the actual findings. On this basis it can be seen that belief in the efficacy of partnership has attained the status of mythology – in the sense that it has become a 'commonly-held belief that is without foundation'.[3]

The Making of a 'Consensus'

First, however, since we are dealing with an apparent consensus, it might be useful to remind ourselves of the profoundly *dissensual* nature of the context in which the partnership model was first espoused in Scotland.[4] This was in the aftermath of the 1987 UK General Election victory of Margaret Thatcher's Conservative Party. However, in Scotland the electorate had roundly rejected her party's neo-liberal agenda. The explanation offered by the Conservatives in Scotland for this outcome was that the Scots had become too dependent on the state and the public sector in too many areas of their lives – that they displayed a 'dependency culture' that made them resistant to the 'good news' of Thatcherism. In turn, Malcolm Rifkind, as Thatcher's Secretary of State for Scotland (responsible for the government's Scottish Office), set out to do nothing less than remould the culture and character of the Scots – to replace dependency and collectivism with enterprise and individualism. The partnership

---

[2] Since the creation of the devolved parliament in 1999 the Scottish Executive functions as the government of Scotland, except in policy areas which remain 'reserved' for the Westminster Parliament.

[3] Chambers Concise Dictionary, Ted Smart, 1988.

[4] For a more developed account of this, see Collins, 1999, pp. 199-204.

model, as outlined in *New Life for Urban Scotland*, was a key part of a broader strategy aimed at bringing this about. This strategy involved, amongst other things, the creation of *Scottish Homes* and *Scottish Enterprise*, the introduction of the poll tax and ultimately the re-organisation of local government itself.[5] *New Life* was to act as a kind of advanced guard for the broader project. Its aim was to tackle the 'dependency culture' where it was at its strongest. This was in the peripheral housing estates or 'schemes' as they are popularly known in Scotland – *bloki* in Polish terms. They were owned almost entirely by local authorities, and had very high levels of unemployment and benefit dependency. The decay in their physical environment was mirrored by a wider process of social decay – family breakdown, crime, vandalism, drug abuse, and so on. In Scotland four such estates were designated as *Partnership Areas* where the government, through its Scottish Office, would lead 10-year pilot projects to establish a new model for the inter-relationship between the public and private sectors in Scottish society. The four areas were Castlemilk in Glasgow; Ferguslie Park in Paisley; Wester Hailes in Edinburgh and Whitfield in Dundee. The model would later be generalised much more widely.

It is in this context that we need to understand the apparently consensual language of 'partnership' in which the *New Life* programme was elaborated. This was made necessary precisely because of the *lack* of consensus about the government's policies, and because of the fundamental legitimacy problems faced by a deeply unpopular (in Scotland) Westminster government. If they wanted to be involved, Labour Party controlled local authorities would be obliged to buy into the idea that they were 'partners', working in harmony with the Conservative controlled Scottish Office. Moreover, any legitimacy problems that central government faced in terms of representative democracy were to be addressed by arguing that the partnership initiatives would be based on *participatory* mechanisms, thus earning their own legitimacy. The 'community' would also be a 'partner'. Thus, the language of 'partnership' as we now know it was born, not out of consensus, but out of a need to manage the profound discord characterising Scottish society and politics during the height of Thatcherism in the later 1980s.

In the period between 1988 and 1998 this *forced* consensus began to take on a 'spontaneous' appearance. Increasingly funding conditions required that everything should be framed to demonstrate the maximum conformity to the Scottish Office's preferred model. And, of course, the best way to do this was to speak the language, and use all the 'buzz words' – as if this was the most natural thing in the world. With

---

[5] *Scottish Homes* was created in 1989. Its mission was to pursue a Thatcherite agenda of breaking up and privatising public sector housing. It was replaced recently by a new body called Communities Scotland. *Scottish Enterprise* took over economic development responsibilities for the country from the *Scottish Development Agency* in 1990. The poll tax (officially titled the 'Community Charge') was a tax introduced in Scotland in 1989 (a year ahead of the rest of Britain) which replaced a tax on property ('rates') with a tax on individual members of the community. It lead to widespread protest and riots and was abandoned by the Major government after the removal of Mrs Thatcher from government in 1990. The Conservatives re-organised local government in 1996 – notably abolishing the larger and more powerful regional authorities in the process.

the Labour defeat at the 1992 General Election, and the resulting generalisation of the partnership model (in the form of the Conservative's 1995 *Programme for Partnership*) this process intensified. Progressively many – indeed far *too* many, including not a few who might have known better – started to speak in the language of the Scottish Office. Increasingly they seemed to do so not just for strategic reasons ('fingers crossed behind the back', so to speak). More and more, it seemed, they spoke this language as if it was their own. The rewards they received for doing this, however, were often very meagre – and not infrequently actually negative (Turok and Hopkins, 1998). Nonetheless, in this way a kind of contemporary conventional wisdom emerged regarding the supposed virtues of the partnership model.

This conventional wisdom shaped the current agenda for local 'partnerships' aiming at 'social inclusion'. Even while the official evaluation of the *New Life* programme was being prepared, the Scottish Office (following the 1997 General Election under New Labour control) felt it safe to say that the programme had been a significant success. While there may have been problems in particular areas:

> The key planks of the New Life approach have ... been widely recognised as the necessary ingredients of effective regeneration (Scottish Office, 1999, para. 7.22).

These 'key planks' were to remain in place in the new *Social Inclusion Partnerships* that were then being designated for the next phase of urban regeneration in Scotland.

'Consensus' and Mythology – The 'Achievements' of *New Life*

One might imagine that for so many people to become so convinced of the merits of the partnership model, something truly dramatic must have taken place. The *New Life* programme must have accomplished something which urban policy has typically failed to do. It must have demonstrated some very substantial achievements, perhaps even have come close to fulfilling some of the exaggerated expectations which always accompany the launch of some new policy or initiative. In reality, this was not so. Instead the *New Life* programme was yet another urban policy failure. The main reason it is not viewed in this way is the unwillingness of the Scottish Office/Scottish Executive, and the many others who have 'bought into' the partnership model, to *allow* it to be seen as such.

Perhaps this reluctance becomes more understandable if we consider the resources allocated to the programme. The consultants commissioned to evaluate the programme report that some £485m of public money was spent in the four *New Life* estates between 1988 and 1998 – around £11,500 per head of population. They also conclude that 62 per cent of this money – some £301m – would not have been spent in these areas had they not been designated as partnership areas (Cambridge Policy Consultants, 1999, pp. 132-133). What, then, did all of this expenditure actually achieve?

A cursory reading of the final evaluation of the programme might lead one to believe that it achieved quite a lot: the consultants do their best to put a positive spin on the programme. They summarise their view of the programme in the following terms.

> As an experiment it has been largely successful, as a programme of public expenditure, it has been value for money. There is a good chance that many of its achievements will be durable. New Life has created a platform that, with good continuation partnerships and fair economic weather, gives residents the chance of a sustainable improvement in their quality of life (ibid, p. 183).

This may seem highly positive. But here again the potential for 'misunderstanding' exists. On a closer reading the passage is actually quite troubling. It suggests in fact that after 10 years and almost half a billion pounds of investment in just four housing estates, there is only a 'chance' that residents of the areas will experience 'a sustainable improvement in their quality of life'. Moreover, this chance, in turn, rests on continuing urban policy interventions and on 'fair economic weather'. This is not at all what the promoters of the *New Life* programme told us to expect. Rather we were told to expect the emergence of stable, well-functioning communities, which would be seen as desirable places to live. So what did the programme *really* achieve?

Housing

It has long been known that improvements to housing stock and the environment are the least difficult part of 'urban regeneration'. This is borne out by the *New Life* evaluation. Between 1988 and 1998 £320m of public money was spent on housing across the four areas – some £18,500 for every household. It is estimated that a further £61m was spent by the private sector (though this required significant public subsidy). And apparently only 37 per cent of this money would have been spent in these areas had they not been designated as partnership areas (ibid, pp. 76-80). Clearly this has made a difference to the housing stock in the four areas, although (perhaps unsurprisingly) little consideration is given in the evaluation report to those other areas that lost out on investment as resources were concentrated in partnership areas. However the housing renewal has by no means been comprehensive, and in two of the four areas – Glasgow's Castlemilk and Paisley's Ferguslie Park – the envisaged renewal programme remained incomplete at the end of the initiatives, and looks likely to remain incomplete for the foreseeable future (ibid, p. 89). The report also makes no mention of the very real problem of negative equity which has affected those who purchased private housing in Ferguslie Park – though such information would have been readily available from owners and local estate agents alike. Given the centrality of 'tenure diversification' to the partnership model, this is a notable omission.

Paradoxically, moreover, the housing programme has actually compounded one of the most worrying trends in each of the four areas – the continuing loss of population. This is a trend that undermines the basis for service provision, and so poses serious concerns for attempts at regeneration. Thus, for instance, Castlemilk based its entire strategy for a 'well functioning suburb' on first stabilising, and then substantially increasing, the local population. In practice it lost a further 25 per cent of its population between 1988 and 1998, and simply abandoned its strategic population targets. It did so, moreover, 'with surprisingly little strategic discussion of the implications' (Stewart et al, 1996, p. 12). Yet, if we look at the findings of the evaluation in relation to housing, we find that across the four partnership areas, demolitions outstripped new build by some 3,133 units (Cambridge Policy Consultants, 1999, p. 81). Thus, the housing investment programmes in the *New Life* partnerships seem to have compounded rather than alleviated one of the most worrying trends across the four areas.

The evaluation's ultimate conclusion on the housing programme concedes that:

> In disadvantaged areas ... housing renewal is a necessary condition for securing regeneration but is by no means sufficient on its own to secure comprehensive economic and social regeneration (ibid, p. 92).

So, given the partial and problematic nature of the housing renewal across the four areas, how did the broader attempts at economic and social regeneration fare? Unfortunately the answer is 'very poorly'.

Labour Market Participation

One of the main aims of *New Life* was to reduce 'dependency' by increasing labour market participation. However the findings show that in two of the four pilot areas labour market participation actually *fell* over the life span of the partnership projects. In Castlemilk it fell from 38 per cent in 1988 to 36 per cent in 1998. In Wester Hailes it fell from 57 per cent to 48 per cent in the same period (ibid, p. 64). In both these areas the 'dependency' problem got worse. In the other two areas the employment rate did improve. However these were the two smallest areas, and the improvements here were associated with substantial movements of population which saw an influx of new groups – most notably owner occupiers with much higher employment rates (ibid, p. 86). In short, improvements in labour market participation across the four partnership areas seem not to have come about through improving the employment prospects of the original population. This view would seem to be supported by the consultants' finding that across the areas there was an 'absence of successful innovation in design and delivery of employment support' (ibid, p. 56).

'Quality of Life' and 'Value for Money'

The evaluation report uses the term 'quality of life' to encompass a 'diverse and subjective' range of aspects of regeneration – including education, crime, poverty, health and community activity. The findings here are truly worrying. For, despite the existence of sustained, high profile, and comparatively very well resourced initiatives: 'the findings do not show large improvements in any of the aspects of the quality of life on the estates' (ibid, p. 123). Readers might by now have guessed that this is likely to be something of an understatement – and they would be correct. In fact the findings show that in a number of important respects there was a significant *deterioration* in the 'quality of life' of the residents of the four estates during the life of the partnership initiatives (ibid, pp. 121-122). For instance, as well as the growing benefit dependency in Castlemilk and Wester Hailes, we find increases in each of the areas in the proportion of residents afraid to leave their home at night. Ironically, this is most marked in the areas where the employment rate has increased (Whitfield and Ferguslie Park). Each of the areas also saw a drop – and in three of the four areas a marked one – in the proportion of residents who attended community, group or club meetings of any kind. This is all the more worrying when one recalls that it was an explicit objective of the *New Life* programme to promote and develop community activity.

It is important to note that the lack of progress on labour market and 'quality of life' issues that we have just reviewed can only cast doubt upon the sustainability of the housing investment in the four areas. In this light, perhaps the most remarkable thing about the final evaluation report is that it argues that these 'achievements' represent 'value for money' (ibid, chapter 6). The basis for this claim is that the programme achieved as much, pound for pound, as comparable initiatives south of the border in England, but that it did so in more difficult circumstances (those of the Scottish peripheral estates). Another view might be that if the value for money achieved by the *New Life* programme was better than in its English counterparts, then at UK level urban policy seems to deliver very poor value for money indeed.

'Partnership Working' and 'Community Participation'

The preceding review of the 'achievements' of the *New Life* programme casts some very serious doubt on the idea that the programme was a success in terms of its substantive outputs. But *New Life* also espoused a very firm view of the kinds of working relationships it wanted to achieve – based on 'partnership working', and 'community participation'. To what extent were these achieved?

Here again a careful reading of the final evaluation is revealing (ibid, chapter 7). The much-heralded idea of partnership between the public and private sectors amounted to very little in practice. The private sector played at best a limited, and probably more accurately a marginal, role in the strategic development of the partnerships (ibid, p. 163). In fact public bodies set the agendas of the 'partnerships', with the Scottish Office playing a leading role. Moreover, even when we consider the relationships among the public bodies involved, the collective nature of the 'partnerships' was 'more apparent than real' (ibid, p. 148). In reality, organisations dealing with housing and employment tended to dominate, and other bodies were marginalized. Important discussions tended to be conducted on a bilateral basis – *outwith* actual meetings of the partnership bodies. And it was generally in these bilateral meetings that important decisions were taken (ibid, pp. 147-148).

It is no surprise then that the issue of community participation proved to be yet more problematic. On this however the final evaluation is strangely laconic – especially considering the volume of research conducted on this theme (and much of it commissioned by the Scottish Office itself). The view seems to be that the less said about the experience of community participation in the *New Life* partnerships, the better (ibid, pp. 153-155). We are told, however, that the community groups in

Castlemilk and Whitfield were *dis*empowered by their involvement in their respective partnerships. We are also told that in Ferguslie Park 'a chasm' emerged between the community and the other partners that has continued since the Partnership was wound up and 'is threatening the future of the estate' (ibid, pp. 155 and 181). Unfortunately, the role of the Ferguslie Park Partnership itself in bringing this situation about is not explored (Collins and Lister, 1996; Collins 2000). Regarding Wester Hailes, a vague reference is made to lack of co-operation between the community and the other partners. No reference is made to the finding of an earlier report, that relations between the community and the other 'partners', 'at a profound level' were not working well (McGregor et al, 1996, p. 23).

In short, the *New Life* programme seems to have been just as unsuccessful (perhaps even more so) in achieving its desired style of working as it was in delivering its substantive aims of social and economic regeneration.

The Continuing Dominance of the 'Partnership' Model

On the basis of the above, the current consensus about the efficacy of the partnership approach to urban policy in Scotland could be viewed as mythology – that is, a belief commonly held but without foundation. For the *New Life* programme, both in terms of substantive outputs and in terms of desired modes of working, has demonstrably failed to achieve its objectives. Indeed, in order to see this, we need only read the final evaluation of the *New Life* programme itself, so long as we are prepared to see beyond the carefully worded 'spin' and the kind of 'misunderstandings' this might create. The actual findings of the evaluation suggest that it is time for the mythology of partnership to be punctured and for a serious debate to take place over the future of urban policy in Scotland. The spin, however, demonstrates quite clearly that this is a debate that New Labour, Scotland's civil servants and others in the policy community, really do not want. That they seem to be succeeding in this is hardly a healthy sign in Scotland's new democracy.

Thus, for instance, the Scottish Urban Regeneration Forum (SURF), a not-for-profit organisation which provides a network of bodies and individuals with an interest in urban regeneration, and which includes among its sponsor members the *Scottish Executive*, Glasgow and Edinburgh city councils, *Scottish Enterprise* and *Communities Scotland*, describe their mission in the following terms:

> To achieve widespread recognition within Scotland that sustainable regeneration can only be realised by adopting an adequately resourced, co-ordinated, multi-sectoral approach.

This is, of course, shorthand for saying that the current model of partnership represents the best possible way of proceeding with urban policy. Moreover, the phrase 'within Scotland' might as well read 'within and beyond Scotland' – since SURF also embraces international activities and comparisons. This shows that a serious and honest debate about the merits of this approach is simply not on the policy-making agenda in 2002. This, combined with the repeated assertion that the current partnership model is 'simply the best' might suggest a striking lack of humility in the policy-making community.

But then the Scottish urban policy community is not 'big' on humility. We have some evidence of this in the form of SURF's creation in recent years of what looks like an equivalent in urban regeneration in Scotland of Hollywood's Oscars' Ceremony. These are the 'awards for best practice: urban re:generation' [sic]. In the context of our appraisal of the actual achievements of the model to which all entries will be expected to conform, this might seem less than entirely appropriate, and perhaps even a little pretentious. It could also be seen as further evidence of the 'immodesty' characterising this policy community. One fears that in reality it has less to do with 'best practice' than with promoting the so-called 'partnership model', as well as the specific sociology, including general careerism, of the urban policy community. We in Scotland may have no choice but to suffer this kind of affliction for the time being. But perhaps it is one of those aspects of the urban policy community in Scotland that will encourage colleagues in Poland and elsewhere to reflect critically upon some of the claims it makes about the 'partnership' model.

'Partnership', 'Participation' and 'Unintended Consequences'

Here in Scotland some of the most worrying implications of the continuing dominance of the partnership model relate to the issue of community participation. On this issue the lessons drawn from the *New Life* programme are of deep concern. The final evaluation argues that this aspect of the partnership model was taken too 'literally' in the programme. It was seen as implying 'a need to get the community involved in the partnership process'. However, while 'this was an admirable objective, it has had a number of drawbacks' (ibid, p. 197). In short, it seemed to give local communities apparently reasonable grounds for disagreeing with the proposals of the other partners, and this too often led to conflict and disagreement, and at times to 'the unreasonable use of relative power' (ibid, p. 163) by local community representatives. The official lesson is that henceforth communities should *not* be encouraged to think of themselves as equal partners in the partnership process (ibid, p. 154). They should be steered away from developing the kind of 'representative skills' that, in the official view, contributed to problems, and encouraged instead to address an apparent deficiency in their 'community development skills' (ibid, p. 162). The closest we come to finding an explanation of what this means is that it will involve equipping local people 'to staff and manage a wide range of projects' (ibid, p.198). What all of this indicates quite clearly is a view that in future the whole process of community participation should be *managed and controlled* much more effectively than was the case in the *New Life* partnership areas. Crucially, community participants should not be seen as community representatives, and the community at large should not expect to have any significant say in the formulation of partnership strategies for their areas. Early evidence from the *Social Inclusion Partnerships* suggests that this is a view that the latest generation of partnerships has been keen indeed to embrace (Foster, 1999, 2000; Velasco, 2000, pp. 16-18). Unfortunately, however, it is a view that is based very much on a consideration of the manageability of local partnerships. It fails to consider the impact that such an approach might have on the dynamics of what are already some quite troubled communities – and here the key lessons of Paisley's Ferguslie Park experience seem simply to have been ignored. Where independent community organisations continue to function in areas of poverty they are like threads holding what is left of the social fabric together. Weakening those threads and replacing them with institutions more convenient for the 'partnership model' should be conducted with an alert eye for 'unintended consequences' (see Collins and Lister, 1996).

References

Atkinson, R. and Moon, G. (1994) *Urban Policy in Britain: The City, the State and the Market*, London: MacMillan.

Boyle, R. (1988) 'Private Sector Urban Regeneration: The Scottish Experience', in Parkinson, M. and Foley, B. (Eds.), *Regenerating the Cities: The UK Crisis and the US Experience*, Manchester: Manchester University Press.

Cambridge Policy Consultants (1999) *An Evaluation of the New Life for Urban Scotland Initiative*, Edinburgh: Scottish Executive Central Research Unit.

Collins, C. and Lister, J. (1996) 'From Social Strategy to Partnership: Ferguslie Park and its Significance for Community Work Practice', *Concept*, 6 (2), pp. 3-7.

Collins, C. (1999) *Language, Ideology and Social Consciousness: Developing a Sociohistorical Approach*, Aldershot: Ashgate.

Collins, C. (2000) *'New Life* in Renfrewshire: Dependency, Enterprise and (Not) Learning', University of Paisley, Faculty of Business Working Papers, Comhairle Series, No. 17.

Foster, J. (1999) 'Overcoming Social Exclusion: Comments on Current Strategies', Paper submitted to the Lord Provost's Commission on Overcoming Social Exclusion in Edinburgh.

Foster, J. (2000) 'Partnerships and Communities' in Velasco ed. (2000).

Gómez, M (1998) 'Reflective Images: The Case of Urban Regeneration in Glasgow and Bilbao', *International Journal of Urban and Regional Research*, 22, 1, pp. 106-121.

Jacquier, C. (1990a) 'Programme of European Exchanges on the Revitalisation of Areas in Crisis: Provisional Report', Commission of the European Communities, Employment, Industrial Relations and Social Affairs, DIV, Interministerial Delegation on Town and Urban Social Development.

Jacquier C. (1990b) *Voyage dans dix quartiers Europeenne en crise*. Paris: L'Harmattan.

McCarthy, J. (2000) 'Urban Regeneration in Scotland: An Agenda for the Scottish Parliament', *Regional Studies*, 33, 6, pp. 559-566.

McCrone, G. (1991) 'Urban Renewal: The Scottish Experience', *Urban Studies*, 28, pp. 919-938.

McGregor, A. et al (1996) 'Wester Hailes', in Scottish Office (1996).

Scottish Office (1988) *New Life for Urban Scotland*, Edinburgh: The Scottish Office.

Scottish Office (1996), *Partnership in the Regeneration of Urban Scotland*, Edinburgh: The Scottish Office.

Scottish Office (1999) *Social Inclusion: Opening the Door to a Better Scotland*, Edinburgh: The Scottish Office.

Stewart, M. et al (1996) 'Castlemilk', in Scottish Office (1996).

Turok, I. and Hopkins, N. (1998) 'Competition and Area Selection in Scotland's New Urban Policy', *Urban Studies*, 35 (11), pp. 2021-2061.

Velasco, I. (2000) *Social Inclusion Partnerships: How Should Local Communities Respond?* Report of Conference held by The Big Issue Foundation Scotland, Glasgow, 31st May 2000.

Chapter 7

Building Regional Development Capacity in Upper Silesia – Problems and Progress

Małgorzata Czornik and Krzysztof Wrana

Introduction

Economic development is a process of change, it takes time, is composed of many activities and engages different social groups, institutions and individuals. Usually it tries to foster entrepreneurship. Change leaders need to have the courage to initiate new initiatives that are sometimes not popular among the local or the regional community. Development success depends in large measure on the ability of those who benefit to convince others of the need for change. The promoters of change need to have a variety of skills, including initiative, determination, abilities in co-ordination, control and persuasion, and a capacity to mobilize financial resources. They also need to be clear-headed, critical and visionary.

Regional development processes should start from identification of broad goals. However it is impossible to create a goals structure for any region without paying careful attention to its specific features, in particular the expectations of its inhabitants. Such expectations are usually very varied and finding overall community-wide agreement can be difficult. Legitimacy is a central issue. The region's development is not dependent on one or a few actors' activity but is a product of the motivations and behaviour of many interest groups and individuals in various alliances. Promotional initiatives can significantly help regional development leaders in achieving the desired results. Sporadic social consultations and goals' negotiation is not enough. Regional development managers (and their results) need to be under constant scrutiny.

Shaping regional development requires an orientation towards the long-term where building a 'static capacity' is less important than 'dynamic capacity' focusing on those factors emphasising learning and adaptation to constantly changing conditions. Short-term success is one thing but the population is usually more interested in systematic, durable improvement in regional living standards.

Local authorities' ability to perceive and build a regional development capacity is an essential aspect in the process of regional restructuring. Such 'development potential' is a crucial resource with strategic significance. It embraces a capacity of local communities to become involved in decision making and policy formulation. The essence of endogenous development is economic and social 'activism' in locating and strengthening an internal potential for development undertakings. Its strategic dimension lies in creating a capacity for a region to connect better with external opportunities and to meet more effectively external competition.

Building such a development capacity in Upper Silesia is a complex problem, with many dimensions engaging a variety of resources and institutions. Upper Silesia's position in Europe is changing fast with many new challenges. The development capacity of the region is based on a number of key elements:

1. Integration of the regional transport infrastructure in a trans-European system.
2. Attracting global market investors.
3. Maintaining a rapid pace of structural change.
4. Strengthening metropolitan functions in the region.
5. Restructuring post-industrial areas.
6. Developing tourist and recreational potential, the natural and the cultural environment.
7. Revitalising human capital.
8. Creating vibrant social capital.
9. Strengthening multicultural identity.
10. Counteracting marginalization.

These factors can be divided into two groups, the 'hard' (1 - 6) and the 'soft' (7-10), and each needs to be nurtured via a different set of policy instruments. Some of them are 'extra-regional', dependent on external decisions and outwith the competence of local authorities. This includes legal, political and financial conditions. Others are entirely the business of local authorities. To build the development capacities of the region Silesian local authorities have developed much in the way of policy, including the *Development strategy of the Śląskie Voivodship 2000-2015* [1] and the *Programme for regional development of the Śląskie Voivodship 2001-2002.* [2] If the intentions of these policy documents were implemented much, undoubtedly, would change in the region. We believe that those visions should be taken seriously for at least two reasons. First, they were created responding to the needs of many external institutions interested and involved in the region, institutions that needed to know the strategic goals of the regional authorities. Second, a majority of the members of the regional community are aware of the necessity of restructuring, the only way to build a new social and economic regional reality.

'Hard' Factors in Development

Integration of the Regional Transport Infrastructure in a Trans-European System

A vital element of regional policy is the elimination of backwardness in the transport system. Existing structures are far from satisfactory. The Silesian voivodship has many roads but almost half are in poor condition. There are no complete ring roads so

---

[1] Sejmik of the Śląskie Voivodship, Katowice 2000.
[2] Sejmik of the Śląskie Voivodship, Katowice 2001.

vehicles must go through town centres. Urban centres in the Upper Silesian Agglomeration have an especially poor roads set-up, made worse by high traffic loads, mining damage and poor quality materials used in road-building. Silesia has two trans-European transport network corridors and three rail routes of the international 'E' system. The motorway building programme is, so far, wholly inadequate in regard to Silesia's needs. The railway also needs modernization. Its track density far exceeds demand. Silesia has its own regional airport, the Katowice airport at Pyrzowice and thanks to new air connections its capacity is steadily increasing. Internal water transport plays a marginal role in the region and this is unlikely to change given that the Odra and Gliwicki canals need large resources and central government support.

Attracting Global Market Investors

Successful restructuring in Silesia needs new industrial investors as well as financial, insurance, science, educational, tourist and trade institutions. All would help change the perception of the region. International investors are interested in Silesia. As is shown in detail in chapter 15 of this volume Silesia is Poland's second (after the Warsaw region, Mazowsze) most important magnet for investment. So far it has attracted at least ten per cent (more than US$3bn) of all foreign direct investments in Poland.

Local authorities attract foreign investors through promotion and the dedicated activity of inward investment agencies. Sometimes such promotional activity takes a multinational, cross-regional character as in Euroregion activities (the *Silesia* and *Śląsk-Cieszyński* Euroregions) or in co-operation with Nord-Pas de Calais in France or Suceava in Romania. Special strategic attention has been given to cross-border integration with the Czech Republic and Slovakia, an important collaboration in preparing and implementing motorway projects and their interconnectivity, an initiative that should increase productivity and effectiveness across all the relevant regional economies.

Maintaining a Rapid Pace of Structural Change

The still mainly industrial profile of Silesia is confirmed by the high level and share of employment in manufacturing, coal-mining and construction. But new visions of the region also exist. The motor industry is important, trade activities have been modernized and services transformed. Faster growth should come from new areas of activity, new products and new enterprises. The priority in development strategy is to increase the innovative potential and competitiveness of the local economy. It is to be achieved through strengthening small and medium enterprises, scientific research, new technologies and improving the general attractiveness of the region. A strategic aim is to support, institutionally and financially, the SME sector, and to encourage technology transfer.

Strengthening Metropolitan Functions in the Region

The development strategy for Silesia aims for 'metropolization' as a process of transformation from a traditional industrial region to an 'intelligent region' where the process of self-learning is established and contributes to new regional features (Senge, 1990). Silesia has a chance of becoming one of the modern European regions. In that process cities have a special role and Silesia has 16 such urban centres concentrated in the central part of the region. They have an important influence on other smaller urban communities and on surrounding rural areas. To acquire a competitive position the region should support the development of its cities. The 'metropolization' of the Katowice Agglomeration is both a necessity and challenge. The situation of the region, the dramatic struggle with its past economic significance creates immense problems. The region's development strategy therefore also includes activities connected with developing metropolitan ambitions. They aim at creating new functions for the principal Silesian towns (especially Katowice). Cities must however be active in shaping their own competitiveness. They should become more heavily involved in marketing an awareness of their economic possibilities.

Restructuring Post-Industrial Areas

One of the conditions in successful restructuring is the availability of attractive locations for new business development. This is a major challenge especially in the central part of Silesia. New business faces serious difficulties in accessing the resources it needs. There are problems with ownership status, the liquidation of existing infrastructure not to mention gaining social approval for new activity. The costs involved in revitalising towns and industries often exceed the financial capacity of the local community. The transformation of post-industrial areas requires building a modern infrastructure, renovation of the urban fabric and, above all, vision and imaginative proposals for land use.

Developing Tourist and Recreational Potential, the Natural and Cultural Environments

The Silesian natural environment is an obvious victim of long years of industrial activity and neglect. But much has happened in recent years resulting in pollution reduction, water quality improvement, afforestation of unused land areas and other activities improving the quality of life and the climate for business enterprises. Silesia has important leisure facilities and potential including the mountainous landscapes of Beskid Śląski and Beskid Żywiecki, the limestone of Jura Krakowsko-Częstochowska, as well as health resorts and forest complexes. The region has eight natural parks and 59 nature sanctuaries. Water sports facilities exist at Jasna Góra in Częstochowa where the shrine of the Black Madonna attracts around 4m visitors each year. Nevertheless Silesia still suffers from a very poor tourist infrastructure (in 1999 it had only 13 hotels, eight motels and 45 bed and breakfast establishments) and a satisfactory level of service could not be guaranteed (although it is changing at a fast pace). There is also a lack of cycle paths but much scope for development.

'Soft' Factors in Development

Revitalising Human Capital

A fundamental condition of competitiveness is that the human capital must exist to sustain a highly effective, good quality innovative economy. In absolute numbers of inhabitants the Silesian voivodship is the second in Poland (almost 4.9m people) but it has the highest population density in the country (397 persons/km^2 as compared to a national average of 124 persons/km^2). The average age is below that of other regions but at the same time the birth-rate is one of the lowest. In Silesia 28 per 1,000 inhabitants (compared to a national average of 33) are involved in study. In spite of this Silesia is one of the most important academic centres in the country. Local government is generally committed to the further development of higher education. Big cities regard research centres as essential to local economic development and regeneration and many co-operate in a variety of ways with new higher education initiatives .

Creating Vibrant Social Capital

A principal feature of Silesia is its urban profile. The voivodship, in 1999 had Poland's highest urbanization rate (79.6 per cent of the population living in towns). Only one in five inhabitants lived in rural areas but even those regions were strongly linked to major urban centres (especially around Katowice itself). This leads to a 'double-employment' effect where many people divide time between urban and rural employment. At the same time the industrial character of the region is apparent in many features of the local community, in relationships between individuals, groups and institutions and in society, economy, culture and religion. Many people have strong attachments, or at least recent memories, of working in coal mines and steel mills. Those economic activities have undoubtedly affected life-styles in profound ways. Much broader national circumstances (before World War II many enterprises were founded by German capital and manned by Polish workers) have also had a strong impact on the region. The results of 40 years of intense socialist industrialization are still perceptible in the first years of the new millennium. Local development strategies increasingly aim to encourage the adaptive abilities of inhabitants, particularly with regard to promotion of entrepreneurship and, in general, the growth of self-confidence.

Strengthening Cultural Identity

Silesian society is deeply rooted and characterized by strong feelings of identity. Its genesis comes from the complicated and stormy history of the region. Its multicultural character and its rich cultural heritage provide a great advantage. For many years the majority of this land was under Czech rule. Before World War I the three great European Empires (Prussia, Russia and Austria-Hungary) shared a common frontier point in a corner of Silesia, at Mysłowice near Sosnowiec and very close to Katowice. It was an area where different cultures, living styles and religions met. Those international connections supported economic activity. Differences were accepted and the region's historical experience supports great social tolerance of religion, customs and behaviour. The region's multinational historical connections, local language, customs and culture can be developed into a strong pro-development factor. The opening of the new Silesian Library in Katowice was one of the most important cultural investments of last few years. The promotion of many different cultural and scientific events is exceptionally significant for regional development.

Counteracting Marginalization

Silesia's economic problems, much associated with the collapse of old industries, have an adverse impact on social relations. A fast growing unemployment rate, increasing poverty and lack of public security fuel a pathology of crime, resulting in marginalization of some parts of society and some city districts. Regional restructuring must improve living conditions giving easier access to social facilities, health and welfare services and other institutions providing social security. Housing redevelopment, including the reconstruction of city centres, is also important. Regional authorities co-operate in such activities with independent groups in 'civil society', with voluntary organizations and others. Overcoming poverty is a difficult and long-term task. Its scale however is such that the engagement of public institutions at various levels (also central) seems unavoidable.

Regional development faces however certain barriers braking the process of improving development potential. This has both internal and external aspects. One internal barrier is a lack of satisfactory information on voivodship potential and resources. Some new institutions should help overcome such obstacles, bodies like the Regional Development Observatory that collects and analyzes information for local and external policy-makers.[3] Another internal obstacle is the slow process of change in people's consciousness, a low faith in their economic abilities, potential and competence. Among those external difficulties the most important are weak and uncertain public finances, a disadvantageous image (as an environmental hazard area with a crisis-generating economic structure) and the fast-growing competitive position of Warsaw, Kraków, Poznań, Wrocław in higher education, in science and in foreign capital inflow. In addition, the lack of real progress with the motorway programme is a serious development barrier in Silesia.

---

[3]'The Regional Development Observatory will be an institution supplying the Board of the Śląskie voivodship with synthetic information about the processes and conditions of the implementation of development strategy. It will collect information and prepare reports.' See *Development Strategy of the Śląskie Voivodship 2000-2015*, Sejmik of the Śląskie Voivodship, Katowice 2000.

Summary

Development capacity needs to be steadily nurtured. It should be concentrated on the need to generate regional competitiveness, new economic activity through restructuring traditional industry and a strong regional specialization around specific economic clusters. The major cities of the region need to ensure good connections with neighbouring urban centres, particularly Kraków, Wrocław and Ostrava in the Czech republic. The core Katowice Agglomeration needs also to assert itself as the centre of a new dynamic of regional development.

References

Development strategy of the Śląskie Voivodship 2000-2015, Sejmik of the Śląskie Voivodship, Katowice 2000.

Programme for regional development of the Śląskie Voivodship 2001-2002, Sejmik of the Śląskie Voivodship, Katowice 2001.

Senge, P. (1990), *The Fifth Discipline. The Art and Practice of The Learning Organization*, Doubleday.

Chapter 8

Social Mobilization and Local Economic Development

Janusz Hryniewicz

Introduction - The Theory of Social Mobilization

The theory of social mobilization deals with analysis of behaviour patterns aiming at the achievement of individual and collective interests. Sometimes it is used to analyse decision making processes to clarify who was most influential and on the basis of which arguments (Tarkowski, 1980). Two broad concepts help explain behaviour patterns leading to the realization of interests: psychological and structural. Psychological concepts stress the personal characteristics or the emotional climate favouring the integration of collective efforts. A typical example of this approach is the analysis of social movements through examination of the emotional character of the ties between its participants (Misztal, 1984). The theory of mobilization has evolved from the structural orientation of social sciences. When we apply it to the collective behaviour of local communities, it is useful to turn attention to the relationships between the theory and 'community development' – understood as an organized activity by members of a given community aimed at satisfying common needs (Wierzbicki, 1983). Jałowiecki (1987) draws attention to two possibilities here, first, a 'top-down' community development when social activities are steered from the top, and, secondly, spontaneous self-organization to exert a 'bottom-up' pressure on local or central governments. The theory of mobilization can be applied to both perspectives. It can also throw light on such matters as:

- the conditions that will either help or hinder the creation of collective initiatives to achieve common interests;
- the necessary scale and depth of mobilization;
- the motivation in participation;
- the chances of success.

The Significance of Social Mobilization for Enterprise and Economic Development

One of the classics of European sociology, Alexis de Tocqueville, researched the stability of political systems of democratic societies and showed that stability is the result of a relatively high collective capacity for self-organization. If politics is about a bottom up competition between programmes and groups then we can expect that the political system will have a relatively large influence on the functioning of the economy. This influence appears through the willingness of members of the community to form various societies and economic associations, and, also through the emergence of a variety of group interests, ideas and programmes for economic development. This is what we understand by 'social mobilization'. Putnam (1995) has shown for example that in Italy regions with greater willingness for social self-organization have achieved greater levels of economic development and their political system is relatively less corrupt.

The link between social mobilization and economic growth hinges on the fact that social mobilization encourages free enterprise and the formation of the technical infrastructure beneficial for business activities, and this in turn directly stimulates economic growth. People's readiness to participate in economic processes is immensely varied. Putman's research suggests that in societies with low mobilization potential relationships will be limited to the family-friendship networks where we see a tremendous disinclination to participate in broader business activity, strengthened by a distrust of all who do not belong to one's own close group. Information circulates within small groups but there is a lack of exposure to different ideas and discursive exchanges. Views on, and connections with, the community at large or its economic problems are not based on any rational analysis, but mostly on emotions and are defensive in character. Distrust means that enterprise is associated with exceptionally high risk because there is no certainty that other people will behave honestly and according to expectations.

Communities where people organize themselves willingly and voluntarily into various societies tend to overcome clan-like isolation. They teach people that others behave according to objective moral norms, broader than the particular norms that rule in their own group. In that way the risks involved in individual enterprise become somewhat subdued.

Societies, and especially political parties and business groups act as sources of new ideas, innovations and improvements of communal life. As they compete with each other they contribute to a specific atmosphere of joint activity within the community showing that collective and individual achievements can bring results. The local 'media' is an important factor in social mobilization. They serve as an important way of overcoming the 'family-friends' partiality by facilitating public discussions on important local problems. In conclusion, we can say that interest groups and local media help in attenuating risk by encouraging enterprise through promoting motivation in seeking individual and collective successes.

Research into factors supporting regional and local development reveals a positive correlation between levels of social mobilization within a local community and local government effectiveness (Hryniewicz, 1998a; Swianiewicz, Dziemianowicz,

Mackiewicz, 2000 p. 38). Social mobilization has the same effect on local government as market competition has on the business enterprise. Large numbers of non-government organizations, relatively strong political parties and local media, all contribute to a climate of 'constant critical analysis'. Continuing and lively political debate means that local government is under constant pressure to review and modify its activity in searching for the most effective methods of administration.

Social Mobilization and Non-Governmental Organizations in Poland

Social mobilization is realized in Poland through membership of political parties, trades unions and non-governmental organizations. The latter include a wide spectrum of societies, committees, clubs, church organizations, self-help societies and various foundations.[1] Social mobilization is highly varied across regions.[2] Looking at the number of organizations per 10,000 inhabitants it varies from 2.8 to 17.8 with an average of 6.1. The Warsaw region is an undisputed leader in social mobilization – with 17.8 organizations for every 10,000 inhabitants. Kraków, with 11.8, has fewer but these two regions clearly lead over all others. The distribution of activity is very skewed and only nine regions report above average indicators. Social mobilization in Poland seems rather low as shown by the fact that in as many as 17 regions the number of organizations (per 10,000 inhabitants) is fewer than four. Low levels of social mobilization are found mainly in East and Central Poland while regions with the highest levels are mainly in West and North of the country. Our own research found a similar spread of activity (Gorzelak, Jalowiecki 1996). The degree of social mobilization found in the central cities of each region is much higher than in small towns and villages in surrounding areas. Social mobilization clearly depends on city size and whether it has regionally significant administrative functions. Non-government organizations tend to locate near centres of regional administration in a dense network offering support and co-operation.

Non-Governmental Organizations and Regional Government

The European Institute for the Regional and Local Development at Warsaw University has investigated institution formation and development at local and regional level especially in light of decentralizing administrative reform. This research notes the changes that have taken place in chambers of commerce, foundations, agricultural bodies, parties and non-governmental organizations as a result of the introduction of the new administrative structure from 1999. The integrating function

---

[1] Research into non-government organizations is carried out in Poland by an independent research institution known as KLON/JAWOR, its information bank has data on more than 32,000 organizations and KLON/JAWOR estimates that there are at least 18,500 active foundations and societies across the country (including local branches). See its website at www.klon.org.pl.

[2] The regions used here are the (49) pre 1999 voivodships.

of regional government was also included in this research, its focus the Marshall's Office (*Urząd Marszalkowski*), which provides support, via subsidies, to organizations and business groups working in a given region.[3]

Agricultural Boards

The Agricultural Board (*Izba Rolnicze*) is a self-governing body that advises the Minister of Agriculture and other central government departments. They also exist in the regions where they advise regional government. The experience of the boards and their success in co-operating with regional government varies considerably region by region. Their success in making an effective transition from old (pre-1999) to new regional structures is also mixed. Overall it seems that the self-government of farmers in Poland, in comparison to other countries, is quite heavily dominated by central administration. The opinions of farming organizations carry little weight in any of the central government organs. In other countries, for example in France, the power exerted by the farmers' self-governing bodies is much greater. There, farmers' organizations manage allocation of agricultural production quotas, distribute subsidies and decide to whom the government should sell land. It is difficult to judge to what extent Polish Agricultural Boards could carry out such tasks. But there is no doubt that an urgent need exists for changes to be introduced in legislation to decentralize the various elements of agricultural policy to the levels of the Agricultural Boards. For example the Boards could have the powers to distribute subsidies and make decisions on land sales (to whom, for what purpose).

Our research suggests that Agricultural Boards are not particularly strong organs for agricultural policy. The greatest barrier to their development is lack of farmer interest. The second barrier is that even the Boards' participants do not have clear ideas as to how the Boards should function. The problem is that a basic ambiguity exists: should they be business groups or a form of self-government. The latter is understood as a body where interests are discussed and synthesized at a relatively general level. Some of the Boards give the impression that they purely represent the interests of the businessmen who serve on their executive committees.

Foundations

The 'foundation' has a status somewhere between a firm and a self-governing association.[4] They are very often engaged in training. A typical example is the Ciechanowie Education Foundation (Fundacja na Rzecz Edukacji w Ciechanowie) which mounts courses for the unemployed. The formal principles governing foundations often give the impression of a collection of contradicting sets of rules, specifically designed so the members would be unable to act as a business but could carry out some economic activities. Forbidding enterprise-type activity means that any

---

[3] Our research, in February-July 2000, was based on interviews and questionnaires. We investigated a range of organizations, excluding those with central functions, in Warsaw, Olsztyn, Zielona Góra, Rzeszów, Krosno, Żary, Olecek, Sierpiec, Ciechanów.

[4] Foundations were defined by legal acts on April 6, 1984 and February 23, 1991.

attempt to direct the activities of the foundation to maximize profit-making is not allowed. Nor is it permissible to share in the success of the organization by drawing benefits in the form of profit sharing among the members of the foundation and its shareholders. The justification for establishing a foundation as a formal organization (with a formally stated purpose and with a formal constitution for the members) is to engage in socially useful projects which for many reasons are not carried out by businesses. Foundations may be created by 'physical' persons, by business institutions and also by public administration. Foundations are maintained either by donations or through the economic activities they carry out. The idea of setting up a foundation may be attractive because they engage in economic activities, can make profits, and in some cases, the profits are tax-free. In such circumstances, local government, by creating a foundation, may be able to meet the needs of a community more cheaply than private enterprise which would be obliged to pay taxes. For the members of such organizations, a declaration that the formal structure is of the foundation type, is profitable, because the economic activity can be carried out in a way that large margins can be added to the charges for services, and since these margins do not have to be treated as profits (in a legal sense), they are therefore tax-free, and can be used to pay wages.

Foundations are the subject of very specific tax-relief laws stating what kinds of activities are tax-free. They include scientific and technical research, culture and sport, environmental protection, charities, health service, social work, rehabilitation of disabled and religious activities. Later, road construction, provision of telecommunications for the countryside and water supply to villages, were added.

Our research shows that it is at the lowest level of administration, *gminy*, (the locality or perhaps parish) that most use is made of foundations. *Gminy* encourage the formation of foundations by giving their support, but the funding usually comes from other sources. For example, in the Gołdap region, the activity of the 'Foundation for the development of Goldapia' is spread across three *gminy*, but its assets come mainly from the Agricultural Property Agency of the state treasury. On the other hand, in Sierpce, the 'Foundation for arts, culture and sport' was created by the *gmina* itself. Quite often the activities of these foundations are similar to the methods used by the Agencies for Development, except that the activities are limited to a local perspective. The foundations we researched had an extremely local character with few links to other levels of administration and no participation in shaping developments of a more regional character.

Business Associations and Regional Administrative Reform

The term 'business association' is an umbrella for a variety of different organizations such as employers' organizations, chambers of commerce, trade associations and so on. All have similar functions. Our research reveals that business associations generally have two types of role: internal and external. The internal function aims at promoting such relationships among members that the board becomes 'a learning organization'. Members exchange experiences and information, helping them to improve their abilities in areas such as planning, organization, sales and international contacts. The external function follows from the fact that they are interest groups, concerned to promote activities recognized as beneficial and desirable by their members from trade fairs to more direct examples of lobbying. Often they exert pressure on the administration to win decisions they find favourable. We carried out a survey, backed by interviews, of members of business associations in the towns mentioned earlier.[5]

Our research indicates that the size of member firms is a significant determinant of the profile of activity of business associations. In this respect they can be divided into three groups. The first consists of those whose members are relatively large firms (usually at least 50 employees). The second group is one where membership is drawn from smaller firms. The third group consists mainly of tradesmen, individuals working in services.

The most common service provided by chambers for members is marketing, organising common participation in national and international events such as fairs and exhibitions. The second advantage of belonging to a chamber is the opportunity to exchange experience (52 chambers said they did this). Chambers also act as local and regional pressure groups (28 acknowledged this). Among other activities provision of training was important and qualifying examinations particularly so for those involved in crafts and services.

Membership gives two kinds of advantages. Firstly, it encourages improved management by learning from others. Secondly, membership increases the power of any one firm and the chances of success in the achieving the firm's aims through collective experience and action. It seems however that the chambers' internal function is by far most important.

A little over half of the existing chambers were involved in setting up new structures to be active across the new (post 1999) regions. There appear to be two ways of meeting the challenges posed by the larger post 1999 regions. First, chambers may decide to merge (or create stable co-operative agreements). Second, the integration of the business community at the regional levels may be achieved by participation in the regional economic parliaments (*sejmiki*). These regional level structures embrace collective organizations and any others dealing with economic problems can also join. *Sejmiki* are new structures in Poland – which are only just being created. Among chamber 'activists' the view predominates that there was no improvement in the readiness of the administrative authorities to cooperate with chambers following the 1999 reform.

We asked our respondents what local authorities should do to utilize better the chambers' potential for regional development. The response was (perhaps not surprisingly), that an important improvement would be a contribution to chambers' finances. However respondents did not necessarily have in mind simple donations but rather the funding of consultancy. Another idea was that there should be invitations to

---

[5] The selection procedure for the research depended on the initial information about whether the association had a regional influence. Branches of associations acting for the whole of Poland were omitted. Some 208 questionnaires were sent out and 60 replies received. Most of the researched economic boards were very small (5 members or less) or small (up to 50 members).

meetings of local authority decision-making bodies as well as the creation of joint committees.

The most important impact of chambers of commerce is that they became the effective centres for the popularization of innovations on a local and a regional scale. This particularly applies to those chambers representing relatively large firms. The smaller the member firms the more important external functions become and the greater the chamber's readiness for politics, taking the form of putting pressure on local and regional government.

Co-operation between local government and economic associations could be extended to engage the intellectual potential of the latter in constructing strategies for regional development. But then we need to beware of the possible 'corporatization' of local government. Enterprise associations are essentially interest and pressure groups. The ambitions of economic associations should not be allowed to conflict with earlier specified strategies of regional development in environmental protection, policies for competition and innovation or social policy. Otherwise, the regional economy could become the stage for anarchic battles for influence where the winning groups are those, who at any given moment, happen to have the best contacts with the executive organs of local and regional governments. The most important aspect in utilizing economic associations for the benefit of economic development is to analyse them carefully and determine which of their activities is of greatest social value. If internal functions (their role as 'learning organizations' and being the local centre for innovation) are the most important then we must consider, yet again, the problem of voluntary or obligatory membership. In case of the decision for the obligatory membership, the boards should take upon themselves wide responsibilities for training, quality, ethics and international marketing.

Political Parties and Associations

Social self-organization is achieved through political parties and non-governmental organizations. The expansion of both types of activity was connected with the change of the political system. Lack of experience in social self-organization meant however that most energy, after 1990, went into creating political parties, as natural offshoots of what took place in the capital (Warsaw) and what was shown on television. Creation of the local branch of a political party is organizationally much easier, especially in a town (or a city) where an office for members of Parliament already exists, than the creation from scratch of another association, which needs to be registered and the activity of which needs to be supported.

Local agencies of political parties, backed by organizational and financial support of the parliamentary offices, began to fill the void. Many associations were organized or taken over by the local party activists. One outcome was that a relatively stable relationship between associations and political parties was created. Of course not all associations exist in the sphere of influence of political parties. Nevertheless, regional and local politics show that subsidies are given to those associations linked in some way to the political party with the leading role in a given local government. In some local communities (*gminy*) relatively stable distributive coalitions were created between political parties and the associations connected with them and, after administrative reform in 1999, those coalitions spread gradually upwards to the levels of districts (*powiat*) and regions (*wojewodztwo*). The importance of political parties was considerably increased as a result of the administrative reform as parties became the distributors of socially desirable goods. The introduction of self-government at district and regional levels meant a considerable increase in amount of money whose circulation and division became a task of local and regional politicians. It was therefore obvious that as a result of administrative reform political parties became socially more attractive.

It can be expected, in effect, that the party political agenda will continue to spread at the local level. An increase in number of new party members may also be expected and will be predominantly motivated by material considerations. On the other hand, an increase in number of party members will, in a natural way, diminish the importance of 'family' (we use the term broadly to refer to membership of common organizations) connections within political parties and will therefore be an important factor in strengthening local democracy. This will follow, as 'family' criteria for the selection of personnel and in processes of decision-making will be replaced by more general rules. Obviously, this will not involve a total elimination of prejudice and of group 'particularism' but it will be an improvement. Political parties are heavily involved in shaping a new regional politics. All parliamentary parties have created regional and district structures. Indeed, among all the institutions surveyed so far (agricultural circles, chambers of commerce and other economic organizations) it is the political parties that most thoroughly achieved an integration of functions and activity across the new vovoidships.

The Integrating Functions of Regional Self-Government

The office of regional governor (*Marszałek*) is, in general, treating as a priority the need to integrate activities across Poland's 16 new regions. One problem, a barrier to this, is the relatively small number of non-governmental organizations that exist that have more than a simply local character and influence. This is made worse by the defensive attitudes of many such organizations, their feelings of being under threat, their dislike of compromise and unwillingness to integrate with other similar organizations. So far some of the best integrating activities carried out by the regional authorities are in sport and tourism and areas where structures have a national character, for example the Polish Red Cross, 'Caritas' and so on. Strategies for regional level integration used by the Marshalls' Offices are, on a whole, properly constructed. Unfortunately, their implementation is made difficult by weaknesses on the side of non-governmental organizations. The most common problems:

* too few non-government organizations;
* concentration of their activities in bigger towns (or cities) with shortage of agencies in small communities (*gminy* and *powiaty*);

- low levels of social self-organization revealing a lack of a sufficient number of people willing to work for the common good;
- lack of trust and apprehension in engaging in cooperation with other similar associations.

This results in many types of socially useful activities being carried out directly by government. This increases the costs involved and may be not so effective as it is more difficult for the administrative apparatus to recognize local needs than it would be for associations with local roots. The objective need for the bureaucracy of local-government to become engaged in a wide range of activities is the result of the low level of social self-organization.

Summary

Changes in the activities of organizations and interest groups, following local government reform in 1999, seem to depend on the degree to which such bodies are connected with public administration. The stronger the connections the bigger the changes that have taken place. In the new regional capitals, agencies for regional development became stronger but in the cities of former regions their role slipped. Among business associations, chambers of commerce consisting of relatively larger firms act as regional centres for innovation, and this activity should be supported by the new administrative structure. The Marshalls' Offices are also playing an important integrating role through their policy of supporting non-governmental organizations. This policy, on the whole, is well constructed but the small number of non-governmental organizations, with their defensive characters, combined with their unwillingness to co-operate and integrate into regional groups, creates a barrier for the effectiveness of such policy. Political parties play an especially important role in the task of regional integration. Parties were the first institutions to build regional structures. On a local level some relatively stable links exist between non-governmental organizations and political parties. Political parties together with some non-governmental organizations are likely to form stable distributive coalitions since when parties gain control over resources they tend to continue to favour the same organizations. However administrative reform has influenced the local political scene positively by enlarging the pool of political activists and by resisting and perhaps overcoming 'family' particularism.

References

Information Bank Klon/Jawor, Information Centre for Non-government Organizations, Warszawa, ul. Flory 9.
Hryniewicz, J.T. (1998), 'Gminy wiejskie w procesach rozwoju gospodarczego Polski' (Local small communities in the processes of economic development of Poland), *Przegląd Socjologiczny*, vol.XLVII/2; p. 189.

Hryniewicz, J.T. (1998a), 'Marketing gminy – innowacja w Zarządzaniu gminą,' (Marketing of small communities – innovations in local government), *Przegląd Organizacji*, nr.3/98.

Gorzelak, G. Jalowiecki, B. (1996), *Koniunktura gospodarcza i mobilizacja społeczna w gminach '95*, (Economic conditions boom and social mobilization in small communities '95), ed. EUROREG, Warszawa.

Jalowiecki, B. (1987), *Lokalizm i rozwój. Szkic z socjologii układów lokalnych*, (Localism and development. Sketch of local sociological systems) Instytut Gospodarki Przestrzennej UW, manuscript.

Misztal, B. (1980), 'Socjologiczna teoria ruchów społecznych', (Sociological theory of social movements), *Studia Socjologiczne*, no.1.

Putnam, R.D. (1995), *Democracy in action* Znak Kraków.

Swianiewicz, P. Dziemianowicz, W. Mackiewicz, M. (2000), *Sprawność instytucjonalna administracji samorządowej w Polsce*, (Institutional efficiency of self-governing administration in Poland), Institute of Market Economy, manuscript.

Tarkowski, A. (1981), 'Autonomia lokalna' (Local Autonomy) in *Władza lokalna a zaspokojenie potzeb: Studium sześciu miast*, (Local government and meeting needs: A study of six cities), IfiS PAN, Warszawa 1981.

Wierzbicki, Z. (1983), Community development in the sociological perspective, *Polish Sociological Bulletin*.

Chapter 9

Society in Transition – Social Aspects in Restructuring Heavy Industry Regions

Marek S. Szczepański

Introduction

In this chapter we focus attention on social and psychological aspects in restructuring Silesian coal-mining. One of the most important elements in reform is rebuilding the consciousness of coal-miners and their families as well as the immediate social environment. In other words, restructuring the branch is in fact about restructuring miners' mind sets, encouraging flexibility and willingness to embrace change. Processes in the more extensive social system, particularly the commune and region, can either contribute to this, hinder or even stop it.

Sociological studies suggest that success in key system transformations depends largely on a positive approach by both individuals and social groups. Individual attitudes may either be positively inclined towards restructuring or indifferent, passive, non-accepting or even violently opposed. If, then, in a social system, in individual or group consciousness, acceptance of change is not established, change will be slow, often resulting in deformations and pathologies. It seems particularly important that the main actors and subjects – regional and local communities as well as individuals – should be aware of the necessity of change.

In what follows we divide our discussion into four parts. First we review recent sectoral transformation and the economic and symbolic degradation to which it leads, touching also on labour market developments. Second we examine how miners adjusted to redundancy and how redundancy payments (from the 'mining social package') were used. In the third part we examine a typical mining commune (Miedźna in the south-west of the Silesian voivodship) whose economy significantly depends on a still working coal-mine, the Czeczott enterprise. Finally, we discuss the wider objectives in restructuring, the rebuilding of the broader economy and its social and political structures.[1]

---

[1] This chapter draws on the author's empirical research, the work of Dr Konrad Tausz from the Central Institute of Mining, the existing literature, including professional publications, the regional and local press and master and doctoral theses prepared in the Institute of Sociology of the University of Silesia.

Economic and Symbolic Degradation and Labour Market Developments

The 1990s brought many positive and negative changes to Upper Silesia. The political silence over mining and metallurgy was interrupted launching a process of transformation. For many years political silence had created a sense of comfort for central and regional government. Sometimes however it is worth disturbing this comfort and not delaying events which must occur sooner or later and which will inevitably destroy workers' illusions. A hibernation that freezes mining's situation is simply bad. The first real attempt to restructure the coal sector was the programme devized by Janusz Steinhoff, the economy minister in the AWS-UW government over 1997-2001. That programme, launched in 1998, proposed job cuts that would cause considerable social, existential and psychological problems and not only for the families of redundant coal-miners.

In the three years, 1998-2000, nearly 85,000 people left mining. In 1998 alone at least 38,300 people left, more than two times the number anticipated in the government programme. More than 10,000 took advantage of redundancy payments worth around zł 44,000 on average (equivalent at that time to almost 20 months average mining wages).[2] In 1999 alone almost 25,000 miners left the sector under the terms of the social package.[3] By the end of 2000 the average redundancy payment was worth zł 55,000 (at that time 17 times the average gross monthly mining wage) but the redundancy target (another 29,000 jobs) looked unlikely to be achieved. By the end of September, 14,400 coal-miners had left their jobs, over half (8,500) taking advantage of lump sum redundancy payments and 'mining holidays'. Despite the slow down in 2000 such a large employment shift in such a short time within one branch probably sets a European record. One should remember also that many miners cultivate a traditional family model in which the father is the only bread-winner. To put it another way, some 85,000 miners maintain 200,000 other people. Assistance for former miners must clearly take the wider social context, and the family, into consideration.

The traditional Silesian mining family experienced immense transformation pressures. As a result of recession, plant closure through bankruptcy or privatization, the number of industrial workplaces in the former Katowice voivodship fell by almost 300,000, only very partially compensated by job creation. There were, and still are, so-called 'social posts' in mining (up to 40,000) and metallurgy (up to 20,000). Those in such posts could leave with no harm done to the functioning of the enterprise but for social and political reasons (lack of alternative places of work, fear of sharply increasing local unemployment and its consequences, concern for political tranquillity) this does not happen.

Economic transformation in Upper Silesia has been accompanied by two types of 'degradation' – economic and symbolic. We turn first to its economic dimension.

---

[2] According to the central statistical office, GUS, in 1998 the average monthly wage in mining was zł 2,253 (US$644) and the industry average zł 1,307 (US$374, both gross) from *Rocznik Statystyczny 1999*, p. 157 and p. 483).

[3] The biggest group took advantage of the mining 'vacation' (15,068) and redundancy payments (9,680), the smallest group of social unemployment benefits (118).

During 'real socialism' miners were a privileged group with average wages exceeding twice the national average. Over 1995-99 things changed and the gap narrowed to 1.3 to 1.7 times the average. A group of very young (43-45 year old) retired miners also appeared on the scene, taking advantage of the right to retire after working underground for 25 years. Their pension is usually half their average working wage but it is worth noting that such young retirees usually have to maintain a non-working partner and will have children at school. In such circumstances pensions do not guarantee a reasonable existence. It is no surprise in this context to read in the press of the sorrow, particularly among the youngest retirees, of those who leave mines under pressure. Many look for new sources of income in activities in the informal sector at the edge of legality. Searching for additional income often leads to the neglect of other activities especially regarding social (and particularly parental) roles in the family.

Economic degradation is also accompanied by prestige-symbolic degradation. It is by no means less important. The ideology of real socialism led to a conviction among the working class that it had a civilising mission and role. One of the recruiting officers who in the 1970s and 1980s scoured the country for youth willing to work in the mines recalled (Hetmańczyk, 1996):

> I was a teacher in a mining school when I ironed my miner's uniform, packed my suitcase and started my journey ... I began to talk about challenges, courage, virility, ... about accommodation, high wages, possibilities of education, free maintenance. After each meeting I could not keep the pupils of seventh and eighth grades away. They asked about the coal-mine and the job. I showed their families a vision of calm, good work. The majority of them realized that if they decided to send a child to Silesia they would not have to spend money on them. What is more they thought that the child would send them money instead.

The job's privileges, mainly for miners, were also well developed. Both employees and mining school students benefited. The transformation of the 1990s brought their partial liquidation, but with no explanation or understanding of what was going on, a deep feeling emerged of being wronged alongside a conviction of international economic plots (for example by the World Bank) aimed at destroying Polish mining.

The restriction or liquidation of economic and prestige-symbolic privilege in mining and steel, alongside the decline in jobs, diminished the attractiveness of the former Katowice voivodship and led to the death of the myth of the regional El Dorado. Its first symptom was a clear, if not dramatic, outflow of local people. In 1991 some 4m people lived in the Katowice region, by 1995 it was only 3.9m. This was equivalent to the loss of a middle-sized town such as Siemianowice or Będzin. It was the outcome not only of a negative balance of migration but also very low, from 1997 negative, natural growth of population (2.0 per 1,000 inhabitants in 1990, 0.1 in 1995 and -0.1 in 1998). Temporary labour hostels were closed and workers returned to family villages and small towns in poor regions in eastern Poland from which they came in the 1980s. Some international migration mainly to Germany also contributed to the population decline.

In this context it is hard not to notice that some features of the traditional worker family appeared to be dysfunctional towards restructuring. In most cases the only breadwinner is the father. Women have few chances in the regional and local labour market because of low educational level and extremely limited job offers. One can suggest, although the thesis requires a thorough empirical verification, that the high participation rate of women in the Łódź region was one of the positive elements of its economic dynamic and the lack of such possibilities in Silesia may become an important obstacle to successful restructuring.

A particular form of existential degradation for the traditional family is the material deterioration of the housing estates. The coal-mines have about 100,000 such dwellings and the branch's difficulties recoil on the living conditions of people housed in the mine's own property, deprived of maintenance, repairs and efficient administration. A similar situation exists in the housing estates belonging to the steel sector. Tenants do not even want to take on flats the steelworks sometimes offer for free.

A large group of people made redundant in traditional industries has appeared and it is a group open to radical and populist rhetoric. Some, particularly in the traditional labour housing estates, or in remaining labour hostels may be the basis of an underclass trapped in a vicious circle of distress, extreme poverty, poor education and dependency on the social security and welfare system. Underclass ghettos exist in many towns, their population the typical clients of local employment offices.

Although there has been much juggling regarding the possible date for Polish EU accession this obsession with the calendar obscures many real problems. For example, assimilating the Polish structure of employment requires radical, almost revolutionary shifts in the workforce. It is estimated that agriculture will cease to be the main source of maintenance and income for 2.8m people, the fuel-energy sector will lose 300,000 employees, traditional heavy industry another 350,000 and light industry 400,000. In this context the consequences of the Polish demographic explosion from the turn of 1970s and 1980s for the country and continental labour market cannot be ignored. In the first years of the new millennium young Polish people will constitute about 40 per cent of all persons searching for work in the whole of central and western Europe. Polish unemployment may increase considerably. Pessimists say another 1.2m will be without work while optimists suggest about 600,000. It should also be noted that over 700,000 Polish people have worked abroad since the year 2000, mainly illegally, not paying social insurance or taxes. Only 200,000 worked legally, mainly in the closest EU countries (Germany, Austria and France). If the most pessimistic forecasts are realized then in 2005 the Polish unemployment total will constitute 3.7m. Silesia is a particularly threatened region because it had the greatest – after the Mazowieckie voivodship – unemployment in the country in absolute terms (256,000 at the end of 2000) although with a relatively moderate unemployment rate (12.3%).

All these processes raise questions about the regional labour market and its possibilities of absorbing those younger workers leaving mining and steel. The extent and capacity of the regional grey economy to absorb labour is also important. Is the regional labour market a bottomless sack or is it already saturated, soon to show a huge increase in unemployment, growing poverty, and a pathology which almost always affected restructuring industrial regions in western Europe in the past? Reflections regarding the future civilizational shape of the region, its educational system and employment structure, are badly needed. The regional labour market and, using a geological metaphor, its social seismic or non-seismic character is dependent on such a vision.

Adjusting to Redundancy

The fate of those who have already accepted redundancy is worth examining closely. We can easily recall sensational press stories about mass tourism of miners to the Caribbean, the empty car showrooms, the excessive drinking and the shot-firers's profligacy. This, and negative redundancy experiences in some European countries led the State Agency for Coal-Mining Restructuring to commission research on redundant miners.[4]

Over the two years 1998-99, in two stages, 1,000 redundant miners were studied. The research was not easy because its focus was on professional and family vicissitudes, including what happened to redundancy payments. So alongside routine research there was a need to look, respecting privacy, at family budgets.

First Wave Redundancies – Job Prospects

The first stage research in 1998-99 showed that more than half of the miners (54 per cent) found a new job. The majority of those who decided to leave mining had already guaranteed themselves new employment. This positive feature, testifying to the miner's providence and discretion, will be less easy to achieve with further waves of redundancy. Those who found employment worked mainly in private firms (30 per cent), much more rarely in state firms (six per cent) or they were the owners, joint-owners, shareholders in family firms (16 per cent). Redundant miners who found new places of work were employed mainly full time (41 per cent) and only a few worked seasonally (seven per cent) or casually (six per cent). Almost 35 per cent of redundant miners were still searching for jobs in 1999 and the main source of labour market information was employment offices, media, relatives and friends. One clear conclusion of our research is that it is indispensable to expand the activities of the Miner's Employment Office. This challenge is particularly important also because nearly one in eight in our sample (almost 12 per cent) had not started searching for new jobs either because of retraining or, as they put it, because of a wish to rest after a lifetime of hard work in the mines.

One positive psychosocial feature among redundant miners is their optimism as regards finding a new job. More than 65 per cent believe that their chances on the labour market are 'sufficient, good or even very good'. Sociology knows that the

---

[4] This research was conducted also with Małgorzata Tyrybon and Andrzej Tomeczek. See also Kicki, W. J. (ed), 1999 and 2000.

greater the individual and group optimism the better the chance on the labour market, in family and personal life. Almost half of miners (49 per cent) succeeded in finding new jobs.

It is important to notice another aspect here, the unequivocal rejection by the majority of miners (71 per cent) of the possibility of leaving Silesia. Some 18 per cent said they would leave the region but even here only under many conditions (involving accommodation, employment and payment guarantees). Migration is relevant only to a very small group of miners, those in the job for the least time and with the highest education. Finally, our research confirmed the progressive vocational emancipation of miners' wives where almost half (47 per cent) stopped fulfilling purely household roles.

First Wave Redundancies – Redundancy Payments

We awaited information concerning redundancy payments with great interest. Press reports of invasions of car showrooms, tourist bureaux and supermarkets appeared to be exaggerated. Former miners tended to place redundancy payments in term bank deposits (31 per cent). A considerable proportion did buy consumer goods such as cars or household appliances or decided to spend vacations abroad (18 per cent). Very often it was the first time such purchases had been made. Sums were also assigned for debt repayment and for simple current household needs (nearly 18 per cent). Some used the money to buy the flats they lived in (four per cent) and one in seven (14 per cent) used the cash for expenses connected with the job planned for the future. As to be expected investments in stock and shares were not so popular. Only three per cent in our sample, mainly those with secondary or higher education, decided to speculate on the stock exchange. One in fourteen (seven per cent) had not started to eat into redundancy payments. Of those who locked up the capital in bank accounts most plan investments connected with jobs (25 per cent) but the second group aimed unequivocally at future consumption (24 per cent).

Second Wave Redundancies – Job Prospects

Our research on redundant miners continued into 1999-2000.[5] The most striking difference between earlier and later research related to persons not searching for new jobs. Recent research shows that more than 33 per cent of redundant miners had not started searching for a new job (including 18 per cent who had no intention to search in the nearest future). Such attitudes were mainly connected with the desire to rest (38 per cent), the intention to take up a job on their own account (29 per cent), or due to age or state of health (nine per cent).

The number succeeding in finding employment outside coal-mining decreased considerably across the research cohorts. As we anticipated, local and regional labour markets became saturated and simple jobs, not requiring high qualifications, fell. It also can be assumed that the number of persons who leave mining and register as

---

[5] My thanks to the director of the research team, Dr K. Tausz, from the Central Institute of Mining for permission to make use of his material.

unemployed will increase slowly although steadily. However at the end of the year 2000 their number did not exceed 5,000 and still remained marginal among the Silesian unemployed. A serious problem is the low level of education of redundant miners. Across our research some 80 per cent of former coal-miners had only either incomplete elementary, full elementary or basic vocational education.

Second Wave Redundancies – Financial Aspects

As might be predicted the financial situation of miner's families, including the group of beneficiaries of the mining social package, deteriorated. Almost 75 per cent had net incomes below zł 500 per family member, a sum close to the social minimum (zł 508). We should emphasize however that information on incomes is based on respondents' own answers and it is known that in labour groups with moderate wages a clear penchant exists to lower declared incomes alongside an excessively critical judgement as to their position in the social pecking order. Nevertheless it is hard to disregard declarations showing that almost 31 per cent of beneficiaries of redundancy payments had family incomes below zł 200 per person. This was the situation of larger families. Even if the subjective self-perception of financial status misrepresents its actual state the fact that more families may be close to acquiring rights to welfare benefits should be taken seriously. As to the material conditions of miners' families, it is worth stressing that almost all redundant miners' households had TV and radio, 19 per cent had computers, 63 per cent cars and one per cent summerhouses. One in five had mobile phones. Redundancy payments were utilized in a similar way as in first round lay-offs:

- bank deposits, purchase of stocks and shares (46 per cent);
- housing (redecoration, purchase, building) (44 per cent);
- current needs or paying off debt (43 per cent);
- car purchase (12 per cent);
- investments in a new enterprise (ten per cent);
- expenses connected with children (education) (nine per cent).

Family net incomes determine how redundancy is used. The lower the income the more often the money was used for current needs and a purchase of a car. Meanwhile the higher income the more often more was invested in a new enterprise or in building a house.

A Case Study of a Mining Community[6]

It is interesting that in Silesia in spite of considerable change and intense threats to livelihoods there are no mass protests, strikes or revolts. Voluntary severance is still the rule and in the case of pit closure employment in neighbouring mines is proposed. However it is hard not to notice that in many cases branch restructuring causes deep frustrations and experiences that are almost traumatic and painful. This occurs mainly in the towns and villages where a coal-mine is often the only employer, the only taxpayer, the source of financial security for the next generations and the centre of cultural life.

In Upper Silesia there are many such places. One example is the commune of Miedźna, deeply connected, almost organically, with the Czeczott mine. This enterprise, initially threatened with liquidation, was merged, mainly for social reasons and contrary to economic rationale, with the nearby Piast mine. In Miedźna there is a lack of alternative employment and the merger, guaranteeing the survival of Czeczott to the year 2005, should give people a chance to prepare, vocationally and psychologically for a life in the future without mining. The most important thing is to use this chance well.

Our research now moves to the micro level. It reports on the Miedźna commune at the start of 2000. Some 600 inhabitants took part including 100 pupils from local secondary schools. And although only one administrative district was studied the outcomes have a general value since many other towns are similarly threatened.

Miedźna is in the south-east of Silesia close to the Wisła, Korzenica and Pszczynka rivers. It has 15,500 inhabitants and an agriculture-mining character. Unemployment in the commune is still not too high in comparison to other Silesian towns but the greatest threat is youth unemployment, those just finishing school. The Czeczott mine is a key employer. Established in the mid 1980s by the mid 1990s the mine employed almost 7,000 but by mid 2000 only around 3,200 people were left, others having departed on the basis of the mining social package. Some 7,300 people live in the commune's main housing estates and it can be assumed that the majority, even at the present level of the mine's employment, is connected directly or indirectly, in earning or by family ties, with the mine.

The social arrangement in the commune is a mosaic with historic villages inhabited by settled Silesians alongside the inhabitants of the mining housing estates, mainly immigrants from other regions. They were brought to Wola and the commune by recruiting officers searching for workers, tempted by chances, very real, of quickly receiving flats and good wages. The villages have an agricultural, and the mining housing estates an industrial, character. The commune's problems are not only to do with branch restructuring, labour shedding at the coal-mine or even the threat of its closure but also social integration. The commune consists of miners and farmers, Silesian and non-Silesians. Established local stereotypes are strong. In the past the farms of Miedźna were rich, large and kept by *pampony,* well-to-do farmers in an old

---

[6] This research was conducted with co-operation of Dr Małgorzata Tyrybon from the Institute of Sociology of University of Silesia in Katowice, Centre of Entrepreneurship J.S.C. in Wola and the representatives of the commune's community.

Silesian language. Recruited miners also had between them clear regional, ethnic and cultural differences. Some came to Wola from the nearby Upper Silesian Industrial Region, others from remote areas of former Częstochowa or from Kielce or Białystok voivodships. These divisions are visible in the commune today and although they do not have a dramatic character they seem to be at the source of some political conflict.

Our study was keen to explore feelings of social optimism and pessimism. The first is connected with other values of civil society such as trust, credibility, openness to change, willingness to engage in social activities. All contribute to success while lack of such values leads to development problems, distortion and deformation, and favours pathological attitudes and social alienation.

In the Miedźna commune some 44 per cent of the population were pessimistic about current conditions (in the year 2000) and 38 per cent expected the year 2001 to be worse still. Only nine per cent were optimistic about the present and 18 per cent thought that the coming years will be better than the last decade. Generally most people (41 per cent) saw the situation in the commune deteriorating and only 14 per cent held the opposing view. Some 45 per cent expressed great apathy. This pessimistic outlook is clearly correlated with self-evaluation of material status. A considerable proportion of people (44 per cent) stated that in their families disposable income was enough only for the cheapest food, rents, electricity, gas, and no more.

As in other similar research younger employees, especially better educated, view their material status and perspectives more optimistically. The worst perceptions are those of old people, the poorly educated and those with the longest job seniority in state-owned enterprises. Disturbingly, almost one in three respondents believed that investment in the education of young people will not enhance their employment chances in the commune. Clearly, basic facts on regional and national unemployment have not penetrated the social consciousness. Among the Silesian unemployed only four per cent have a higher education and the majority of the long-term unemployed had completed only basic schooling.

Our research also revealed how individuals viewed education in more general terms. It is evident that in the social consciousness, especially in small local communities, a view exists that it is indispensable quickly to finish school and gain a good and specific profession that means regular paid wages. Such views are typical of Upper Silesia, reflecting deep habits and ways of thinking (although under erosion). Over the decades work in a mine guaranteed a well paid job and fringe benefits. In the case of Miedźna supporting full time studies is perceived and treated as one of the best investments that a family can make for its own children, a good protection against unemployment and its individual and social consequences. More and more parents, even the most poorly educated, want a higher education for their children, especially their daughters. The last attitude is at least indirectly connected with a social puzzling fact that education is beginning to be treated as a kind of good marriage asset for a daughter and life equipment for the son.

For the moment parents' decisions regarding education for their children are still not the best informed. Younger parents prefer secondary technical schools, but all this guarantees, often, is education for a surplus profession, doomed for unemployment. Others opt for grammar schools giving good manners and access to university.

A striking and unexpected feature in our research concerns fears and hopes regarding alternative employment possibilities. It is amazing that almost all (94 per cent) expected local government to create new places of work and restrain unemployment growth, although it should be pointed out that our research was conducted at a critical moment, when the mine's closure seemed imminent, before the five year extension it later won. In 2000 unemployment was not high but the threat of its growth was very real. The commune had no clear post-industrial strategy and our research shows that people expect much more from local authorities. People are convinced that the place where they live has a strong influence on employment chances and that they are much greater in urban areas than in small communes like Miedźna.

Building an alternative labour market, rejecting the spectre of mass unemployment on a local scale is partially connected with the expansion of the small and medium-sized enterprise sector. A majority of our respondents (61 per cent) believed that such entrepreneurship could be developed in Miedźna. However only one per cent would opt to set up on their own and 37 per cent declared that they would not go for this kind of the activity.

Much can be done to promote a more positive image of the commune and its perspective. In the past 15 years it was defined as a part of the mining landscape with its associated housing estate. In the transformation decade of the 1990s the conviction became established in Upper Silesia that the Miedźna commune, or more precisely the mining housing estates located in it, were black investments, with poor economic and social prospects. This kind of thinking needs to be changed and a start would be to challenge local information policy. Our respondents very much wanted to see such a change with positive promotion of the commune, most (48 per cent) believed after all that the commune was an attractive place for future investors.

Wherever radical economic and social transformations occur there is an increase in declared readiness to leave the locality, region or even the country. This is the product of ambivalent feelings. On one hand declarations of mental and physical mobility are consoling to individuals. Such local and regional migrations have taken place in recent times in American towns dominated by traditional industries, in mining communities in France and Great Britain. On the other hand the constant readiness to depart brings dangerous consequences resulting in social ossification or even the death of the village or commune. This has happened in many places, especially in the former state-owned farms located in the north-west of Poland. Moreover this attitude weakens attachment to the local community and restricts more positive action in favour of development of the immediate locality. In Miedźna over half in our sample declared a readiness to leave. In the case of young people the proportion was even greater.

As might be expected the greatest proportion of those saying they might depart were living in the mining housing estates of Wola. Many of its inhabitants came to the commune only a few years ago and they do not feel so connected with the local environment, its problems and society. For some it is not Wola and its blocks but the villages and towns they lived in before, and where they left their relatives, that are a positive frame of reference. Those who have the possibility of return consider it seriously. But in reality most will stay and the problem of mining redundancy needs to be solved in the region. The expectations of some that former miners will settle in towns left by the Russian army, or leave for the East, or become Alaskan miners are completely unjustified. It is an irresponsible illusion that the mining problem can be exported outside the boundaries of the region and the voivodship.

What can local and regional authorities do to try to promote a new development strategy? First, much better information policy is crucial adjusted to local culture and competences. People need to be convinced that real development possibilities exist. Local information activities should be directed to create and support the development of civil society. Tolerance towards cultural difference is of key importance. Silesian and non-Silesian inhabitants have much to offer. Cultural festivals celebrating diversity need to be encouraged. The miners versus farmers football match was a small, but important event in this sense. But the labour market and employment is the most important feature. One of our respondents recalled that the region's strong assets were potato fields, bison and forests. Why not try then to make Miedźna famous not only for the Czeczott mine but also for fries, chips or potatoes? Of course, this type of production does not generate high quality jobs but it does something. Beautiful forests and the proximity of bison reserves offer a superb chance to create a tourist and recreational enclave.

Much more could be done for young people. Education needs to be substantially restructured. Sadly, many existing schools educate young people only for unemployment restricting their professional chances in modern sectors of economy. Given the respect it enjoys the church can also play a role here. Poor but capable students who want to continue into higher education need far greater support. Finally, a broad marketing effort is needed, embracing all the important institutional actors.

Wider Objectives in Regional Restructuring

As has been stated earlier, it is short-sighted to separate the problems of hard coal mining and other traditional industry from a wider regional and national context. Sector transformation may lead to favourable changes within the branch itself but the environment matters too. Mining and traditional industries' reform should be correlated with wider activities on the regional and national scale. Strategic objectives of regional development should involve:

1. Stimulation of the regional economy on the basis of a network of innovative enterprises, supported by entrepreneurship incubators, agencies for regional development, information and entrepreneurship centres, technology and science parks. Success depends to a large extent on the involvement of foreign capital.
2. Further reforms of state-owned industry, in particular mining, steel and rail.
3. Privatization of most large and medium state-owned industrial plants, including coal mines and steelworks.
4. A fundamental increase in services sector employment.
5. Creation of greatly improved institutional retraining possibilities. The children of those workers from restructured sectors need special attention, the right education choices are very important.

6. Guaranteeing basic social care for the unemployed and their families. This is especially important given the forecast of a qualitative increase in unemployment in some Silesian districts.
7. Consolidation of regional and local political, financial and technocratic elites on the basis of a fundamentally modernized educational system.
8. Construction of a new regional and cultural identity on the basis of preserved yet diverse cultural cores.
9. Development of architectural and urban space of the region.
10. Environmental improvements.

In constructing a regional restructuring plan one must not overlook the achievements of other countries and regions. Polish experiences should not be the only frame of reference since they are tightly connected with backwardness of the country and its peripheral economic location in Europe. The most developed European countries and their regions have already entered the post-industrial (information) society whereas Poland and Silesia in particular, is still in an industrial phase. The main feature of post-industrial society is the domination of the service sector, not only services in the traditional sense (trade, crafts, transport, recreation, health care) but also, and perhaps most of all, modern services (in business, computing, education, banking, scientific research, telecommunication, real estate, insurance and medical services). In post-industrial countries services employment amounts to some 70-80 per cent of the total. In Silesia that index amounted to only 43 per cent in 1999, testifying to the scale of 'deformity'. It was also unfavourable even in comparison to the national picture where, in 1997, 48 per cent of employment was in services. In the major cities the proportion was 61 per cent in Warsaw, 56 per cent in Gdańsk, 53 per cent in Poznań and Kraków and 52 per cent in Wrocław. In Silesia however over half of all employees (51 per cent in 1997) are employed in industry, the majority of them in mining, metallurgy, and basic chemical industry. Therefore, it is important to fix, in the regional consciousness a conviction regarding the necessity of radically increasing employment in services. This is not made easy mainly due to previous ideological and pedagogical manipulation, the conviction of 'the mission of the industrial working class' and the purposeful promotion of development of heavy industry.

Restructuring Upper Silesia requires the mobilization and co-operation of numerous partners in various fields and various levels. It is necessary to engage central government and its budget. This needs to be supplemented by the conceptual, financial and organizational readiness of the regional establishment, local governments, enterprise management and trade unions. It is crucial that employees, both as groups and individuals, as well as members of their families, are also ready to accept change. Those conditions have not been fully met so far and regional trade unions definitely exclude radical yet indispensable reforms. They reject all forms of collective redundancies, reductions of subsidies, and prompt liquidation of unprofitable mines.

Restructuring is always performed for somebody and in somebody's interest, simultaneously infringing interests of individuals or social groups attached to the *status quo* or even fighting to restore the *status quo ante*. Therefore, this is a game to break even, in which one man's success is another man's loss. In some extreme cases, this may prove a game in which the result is negative, in which – at least for some time – all participants turn out to be defeated. Passive or active resistance towards changes is most of all manifested by individuals or social groups for whom modernization appears to be a threat to livelihoods. Those in the fray tend to have poor qualifications.

Employees and whole local communities cannot be taken by surprise, and must be properly prepared for change. They should not as has been a rule so far, get information about their own destiny from the press, incidental information or gossip. Restructuring should not be associated with threats to existence. On the contrary, it must create new prospects clearly linked to individuals and individual efforts. Mistakes in information policy, lack of clear programmes and restructuring principles are very costly. In the past workers very often were convinced that making people redundant was not due to economic reasons but was the result of political manipulation of superiors, ministers and central government. Although it would certainly be an exaggeration to describe the present situation in the labour market as panic, one should not underestimate the overpowering fears of employees fearing redundancies. If information policy is not radically improved, the dispossessed may become the clients of radical and relatively well-organized political parties and forces.

References

Hetmańczyk, K. 'Najemnicy. Te czasy już nie wrócą' (Hired Hands. Those times will not return), *Dziennik Zachodni*, No. 212, October 25, 1996.

Kicki, W.J. ed. (1999), *Socjologiczne aspekty restrukturyzacji przemysłu węgla kamiennego na Górnym Śląsku. Losy zawodowe pracowników kopalń odchodzących z pracy z wykorzystaniem jednorazowych odpraw pieniężnych.* (Sociological aspects of restructuring the hard coal industry in Upper Silesia. The fate of mining employees departing voluntarily with redundancy lump sum payments), (Szkoła eksploatacji podziemnej 1999, Polska Akademia Nauk Akademia Górniczo-Hutnicza), Kraków 1999, pp. 581-593.

Kicki, W.J. ed. (2000), *Socjologiczne aspekty restrukturyzacji przemysłu węgla kamiennego na Górnym Śląsku. Losy zawodowe pracowników kopalń odchodzących z pracy z wykorzystaniem jednorazowych odpraw pieniężnych. Relacja z drugiego etapu badań.* (Sociological aspects of restructuring the hard coal industry in Upper Silesia. The fate of mining employees departing voluntarily with redundancy lump sum payments. Report on the second stage of research), *Wiadomości Górnicze*, No. 1, 2000.

Rocznik Statystyczny 1999, 2000 (GUS, Warszawa, 1999).

PART IV

STIMULATING ENTERPRISE

Chapter 10

SMEs and Regeneration: A Comparison between Scotland and Poland

Mike Danson, Ewa Helińska-Hughes and Geoff Whittam[1]

Introduction

Across the developed world, but particularly within Central and Eastern Europe, old industrial regions are facing the need to restructure their economies to address the decline of their traditional industries. For some areas of the European Union, this process has been underway for several decades with programmes and policies to attract inward investment and to introduce new and small enterprises to replace the dominant integrated plants of the coal, steel, shipbuilding and heavy engineering sectors of the past. New institutions such as regional development agencies have been established to deliver those policies and they have been working progressively more in partnership with other economic players.

Recent interventions in Scotland, which has often pioneered innovations in regional development strategies, have sought to organize those institutions and partnerships into more coherent arrangements for policy delivery. This paper describes the position in Scotland and the reasons for changes in its economic development process. It is hoped that our description of policy development in Scotland will not only be useful to those developing similar policy superstructures elsewhere but may also stimulate wider debate on appropriate programmes for the early 21st century.

We begin by introducing the theoretical underpinnings of the policy framework in Scotland, concentrating on endogenous approaches that stress the importance of support for indigenous development. The key role of regional development agencies as the vehicles to deliver many of the policies consistent with this paradigm is discussed. We move then to examine Scottish strategies and institutional arrangements. Given the significance attached to small and medium sized enterprises in restructuring, the strategy to improve the numbers of new businesses in Scotland is analysed. This is followed by a review of a recent

[1] This chapter draws on interviews in various institutions in Poland (Warsaw and Katowice) in February 2002. The authors take this opportunity to thank those who provided time, material and helpful insights into the functioning of the Polish SME sector, particularly Mrs Mirosława Płyta from the Polish Agency for Enterprise Development, Mrs Iwona Czaplikowska and Grzegorz Giemza from the Upper Silesian Development Agency.

Scottish Parliament inquiry into local economic development services for new firms, and the Scottish Government's response to the research findings. We then shift focus to Poland and Upper Silesia (the Śląskie voivodship) the Polish region which, with its concentration of heavy industry, is facing restructuring problems that Scotland had to deal with in the not so distant past. We describe the Polish SME sector and its institutional support framework, paying particular attention to developments in the Upper Silesian region. Poland and Silesia have adopted regional development agencies, partnerships with an emphasis on small and medium enterprises that is very familiar. We speculate that the Scottish experience may be interesting and useful to policy-makers in Polish industrial regions undergoing restructuring. The conclusion explores some of these issues.

Endogenous Development and the Role of Governance Structures

In this section the paper focuses on the theoretical framework which has informed policy in the field of regional economic development and regeneration within industrialized economies. It is argued that the organizational structures which are established to reap the perceived benefits highlighted by the theoretical analysis become the key element for success at the regional level. Organization, in other words, is critically important to successful regional economic development.

Over the last two decades, growth economists have identified knowledge and learning as the key feature of endogenous growth theory. This has also been developed by academics and practitioners in regional science: the ability to learn being identified as a key element in achieving a competitive advantage for local and, in some instances, national economies. Likewise, the role of knowledge and the innovation process has been identified as being a significant factor in endogenous growth, and, technically, in accounting for the 'Solow residual'.[2] The key component of the innovative process is the creation of knowledge, a point recognized by theorists and practitioners alike (Gregerson and Johnson, 1997, pp. 479-490). The recognition of this point has focused attention on how firms, regions and indeed nations innovate. In researching this phenomenon, regional scientists have utilized the methodology of the evolutionary economists.[3] A major contribution to the discovery and utilization of knowledge is attributed to the institutional framework, which evolves to facilitate the process. Institutions in this sense include both formal organizations, such as regional development agencies, and informal. The latter can be defined as 'a social organization which, through

---

[2] The unexplained factors contributing to growth and attributed to technological change. The literature is vast see: Boltho and Holtham, 1992, pp. 1-14; Van der Ploeg and Tang, 1992, pp. 15-28; Scott, 1992, pp. 29-42.
[3] There is a very wide literature here see Gregerson and Johnson 1997; Morgan, K. 1997, pp.491-503; Grabher and Stark, 1997, pp. 533-544; Storper, M. and Scott, A. J. 1995, pp. 505-526; Storper, M. 1992, pp. 60-93; Storper, M. 1995, pp.191-221; Lundvall, B. A. and Johnson, B. 1994, pp.23-42.

the operation of tradition, custom or legal constraint, tends to create durable and routinized patterns of behaviour' (Hodgson, 1988).

The emphasis from the regional science perspective is that due to the nature of knowledge – much of it being *tacit*, it is not easily transferable – the system or region that creates it will develop the advantages associated with innovation. Furthermore, Storper highlights the importance of 'untraded interdependencies', which arise, he argues, from the way technology is developed. Utilizing the evolutionary approach he focuses on the path dependency nature of technological development: 'Technologies, for one thing, are subject to a variety of user-producer and user-user interactions.' (Storper, 1995, p. 204). Where clustering occurs because of some commonality of technological development then 'untraded inter-dependencies' arise, such as common coded language, norms, customs and practices. These common institutions lead to easier communication and facilitate trust and co-operation. Similarly, Freeman (1994, pp.463-514) comments,

> Firms learn both from their own experience of design, development, production and marketing and from a wide variety of external sources at home and abroad – their customers, their suppliers, their contractors ... and from many other organizations – universities, government laboratories and agencies, consultants, licensors, licensees and others.

Following the collapse of many old industrial regions over the last two or three decades, states at all levels have tended to promote local and regional economic restructuring based on foreign direct investment, new firm formation and the growth of small and medium enterprises (Steiner, 1984). This has led to the proliferation of policies, programmes and initiatives for such regions and a crowded landscape of agencies and authorities to prosecute their delivery (Danson, Halkier and Damborg, 1998). In the face of a myriad of policy regimes and structures, a number of theoretical approaches have been adopted to explain the implications of such arrangements for the effectiveness and efficiency of economic development in these areas (Cameron and Danson, 2000). There has been a particular concern with institutional thickness and capacity stressing the significance of the delivery mechanisms themselves (Amin and Thrift, 1994, pp. 1-22). Much research has addressed the emerging issues of how evolving regional governance systems are related to this institutional capacity, with interest in regional development, institutional and evolutionary economics, fiscal federalism, and the devolution.[4]

The Regional Development Agency (RDA), established as a specialist institution dedicated to delivery of programmes and region-specific policies, is at the heart of new directions in regional economic policy, Halkier and Danson's (1997, pp. 241-254) typology of the RDA showed that the control and co-ordination of local and regional economic development powers and functions was becoming an important RDA function. In a comprehensive analysis of the development agency as a semi-

---

[4] Once again an extensive literature exists. See Armstrong, H. 1997; Amin, A. and Thrift, N. op. cit, D. Newlands and J. McCarthy, 1999; Wannop, U. (1995); Heinelt, H. and Smith, R. (1996), Hooghe, L. (1996).

autonomous institution Danson, Halkier and Damborg, (1998) concluded that RDAs were beginning to evolve into organizations not aiming to deliver economic development services directly as much as to co-ordinate such activities. Acting, in this way, as catalysts RDAs were already moving beyond being bodies engaged themselves in development in favour of working in partnership with other players in the regional economy.

On a micro-level, this relationship between individual development bodies and their environment has been considered in more depth since then (Danson, Halkier and Cameron, 2000). Building strong relationships between and across public and private sector organizations undoubtedly has been seen as a priority for public and semi-public development agencies in the last few years.

On a meso-level, the role of development agencies in the political make-up of their region is more about networking and regional governance. The increasing reliance on individual agencies or networks of development bodies raises important questions about the relationship between functional efficiency and democratic legitimacy, not least when the European dimension in the form of structural funds programmes is involved (Roberts, 1997).

Finally, on the macro-level, the relationship between regionally-based initiatives and the emerging European system of governance in Western Europe has been discussed in terms of regional development and multi-level governance (Armstrong, 1997). Over the last few decades the traditional regional subsidy programmes of central government has become much diminished alongside a veritable explosion of bottom-up development bodies and initiatives. The structural funds mechanism has linked together these two aspects in an intricate pattern spanning Europe. In trying to understand new developments on the regional and European levels, the transformed role of central government – and therefore the relationship between the national, regional and European levels of governance – is critical. In the following section we address the development of policy in Scotland, which has often been the cauldron for the evolution of new approaches to promoting and implementing regional economic strategies (Danson, Fairley, Lloyd and Turok, 1999, pp. 23-40).

Review of Scottish Strategies and Institutional Arrangements

The list of players which firms interact with, as identified by Freeman is not unlike the institutions identified by *Scottish Enterprise* (SE) as making up a typical cluster: companies, customers, suppliers, utilities, research institutes, education.[5] The SE 'cluster strategy' is not only promoted as a key element of its overall approach to economic development but is also integrated through a number of other strategic initiatives. The recent launch of the 'Know-How' human capital strategy claims in particular to address the need to nurture education, training and learning more generally. In establishing both a Minister and a Committee of

---

[5] http://www.scotent.co.uk

Enterprise and Lifelong Learning, the Scottish Executive and Parliament seem to recognize the critical role of endogenous learning based along indigenous lines.

The Business Birth Rate Strategy (BBRS), launched by *Scottish Enterprise* in 1993, certainly underpins this point. Its objective was to close the gap with the rest of the UK in terms of business start-up, that is, to promote the birth rate of indigenous companies. This was to be achieved by increasing the numbers of businesses started, ensuring more of them survive and more achieved significant growth. Along with measures one would expect, such as improving access to finance, were several other initiatives such as 'building enterprise into the education system', making a very real link between formal education and the BBRS, and highlighting the concept of 'Learning Economies' outlined earlier.

It must be remembered that the BBRS was initiated by then newly-formed Scottish Enterprise Network. While the BBRS has been critically reviewed elsewhere *Scottish Enterprise* has recently undertaken its own evaluation.[6] The Scottish Enterprise review is not encouraging,

> Between 1981 and 1997, the number of VAT-registered businesses per 10,000 population in Scotland increased by 17,000. But there has been little change in Scotland's relative position: in 1980, Scotland stood at 83.7 per cent of the UK average. In 1997, the figure was 84.4 per cent (*Scottish Enterprise*, 2000, p.16).

On the other hand, removing the most prosperous region, the South East of England, meant that 'the number of businesses per head of population increased from 86 per cent of the UK average in 1981 to 91.6 per cent in 1997' (*Scottish Enterprise*, 2000, p.16). Despite any statistical ambiguity, depending on how the data are presented, *Scottish Enterprise*, on the positive side argues,

> The Business Birth Rate Inquiry initiated a nation-wide debate about entrepreneurship, and changed thinking about entrepreneurship where it matters most – within the media, the educational system, the finance community and among policy makers (*Scottish Enterprise*, 2000, p. 6).

Scottish Enterprise certainly believes that the BBRS was successful in placing the creation of an entrepreneurial culture on the agenda of the organizations that help to shape and develop the economic environment.

Another recent study which will impact on the delivery of support to SMEs throughout Scotland is the *Inquiry into the Delivery of Local Economic Development Services in Scotland* of the Enterprise and Lifelong Learning Committee (ELLC) of the Scottish Parliament. The specific remit of the inquiry was to establish whether there was duplication and overlap between various institutions in the delivery of business support services. Essentially this involved a baseline study to establish how existing policies such as the BBRS were implemented in practice. This research (conducted by the Paisley Business

---

[6] For a critical review see Whittam, G. and Kirk, K. 1996, Scottish Enterprise's own evaluation can be found in Scottish Enterprise 2000, full details in References.

School) involved a 'mapping exercise' which highlighted the local development services based on 18 areas across Scotland; an examination of the main generic issues involved in business support; a table identifying the support available and the agencies responsible for policy delivery. Figure 10.1 shows how institutions co-operate to deliver business development services in a typical area of Scotland.

While the inquiry sought to identify what was actually being delivered, the research identified two other non-tangible issues. First a particular form of support landscape exists, with the partnership approach dominating over single agency actions and this helps to some extent to overcome duplication and overlap in Scotland. Second, a rich array of community, voluntary and trades union bodies interact to offer many support mechanisms that often go unnoticed.

The inquiry came to three main conclusions. First, notwithstanding the existence of many useful partnerships,

> There is congestion within the field of local economic development in Scotland. There is confusion, overlap, duplication and even active competition between the many agencies involved (ELLC, 2000, p. 4).

Second, there is good practice of partnership working at the local level, but, third, this in itself 'is unlikely to deliver a level of rationalization of services, cost-effectiveness and consumer focus that is desirable'.

Following from the Scottish Parliament ELLC report, the Scottish Government adopted many of the recommendations of the Committee and incorporated the need to improve co-ordination and delivery of policy into its future strategic framework for the economy. Labelled 'Framework for Economic Development in Scotland' (FEDS, Scottish Executive, 2000) this suggested for the first time a coherent approach to economic growth and development, with special attention to the avoidance of duplication and overlap in the institutional framework. Within this national policy framework, a need was perceived for greater clarity on how local economic strategies were developed and implemented. While in the 1980s economic geography focused on the problems of underdeveloped institutional capacity (Kafkalis and Thoidou, 2000; Amin and Thrift, 1994), the dominant concern of both Parliament and Government in the first decade of the new millennium came to be a preoccupation that there might be too many support mechanisms and organizations. One of the conclusions in the ELLC report was that a local economic forum (LEF) should be established in each area in Scotland to promote a simpler, more cohesive structure for local economic development. The Scottish Government Minister responsible introduced the concept of LEFs in these terms, and suggested their role should be 'to agree a local shared vision and programme of action for the streamlining and improvement of service delivery.' (Scottish Executive, 2000, p. 3).

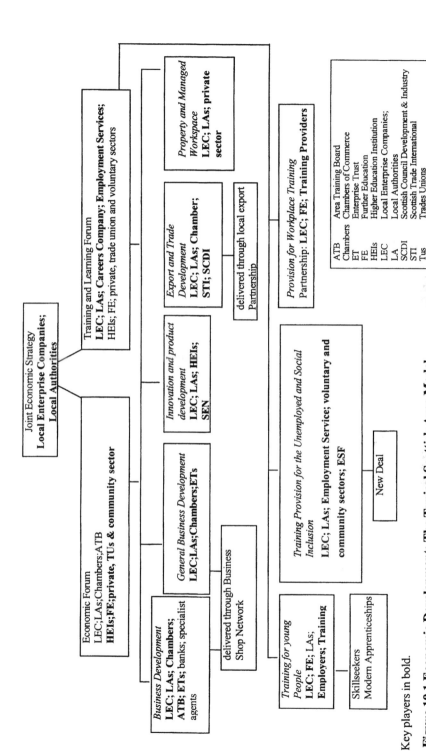

Figure 10.1 Economic Development: The Typical Scottish Area Model

Joint Economic Strategy
**Local Enterprise Companies;
Local Authorities**

Economic Forum
LEC;LAs;Chambers;ATB
HEIs;FE;private, TUs & community sector

Training and Learning Forum
LEC; LAs; Careers Company; Employment Services;
HEIs; FE; private, trade union and voluntary sectors

Business Development
LEC; LAs; Chambers;
ATB; ETs; banks; specialist
agents

delivered through Business
Shop Network

General Business Development
LEC;LAs;Chambers;ETs

*Innovation and product
development*
**LEC; LAs; HEIs;
SEN**

*Export and Trade
Development*
**LEC; LAs; Chamber;
STI; SCDI**

delivered through local export
Partnership

*Property and Managed
Workspace*
**LEC; LAs; private
sector**

*Training for young
People*
**LEC; FE; LAs;
Employers; Training**

Skillseekers
Modern Apprenticeships

*Training Provision for the Unemployed and Social
Inclusion*
**LEC; LAs; Employment Service; voluntary and
community sectors; ESF**

New Deal

Provision for Workplace Training
Partnership: **LEC; FE; Training Providers**

ATB	Area Training Board
Chambers	Chambers of Commerce
ET	Enterprise Trust
FE	Further Education
HEIs	Higher Education Institution
LEC	Local Enterprise Companies;
LA	Local Authorities
SCDI	Scottish Council Development & Industry
STI	Scottish Trade International
Tus	Trades Unions

Key players in bold.

The main initial task of the LEFs has been set out as the eradication of wasteful duplication of business support services and the enhancement of overall service delivery. Complementing other initiatives, aimed at removing confusion and overlap in small business support, the LEFs are to achieve further streamlining. 'Partnership overload and overlap' is also seen as an obstacle to effective and efficient local economic development, so the forums will also be required to rationalize the number of local partnerships. Their coordinating role will be critical in their success, with each forum also a partnership of agencies active in the local field.

Value for money, and especially using public sector resources efficiently, is the key driver in this initiative according to the consultation paper. It is suggested that this initiative should be undertaken within certain parameters: the shared vision and action plan for their area should consider the most effective way of engaging private and social economy businesses in the design and delivery of services. It is assumed that the LEFs will have small, focused cores of local authorities, local enterprise companies (the local, state economic development agencies), businesses, the local learning sector – further and higher education institutions, and tourism agencies. With a tight core membership of about ten, wider consultative mechanisms will include other players and partners: trades unions, housing agencies, national training organizations, and others.

The tasks set out for the LEFs in their first year (2001-02) include a comprehensive mapping exercise of the delivery of services to businesses, an evaluation of the quality and effectiveness of economic development services, and a review of existing partnership structures to assess the need for them in the ever more crowded policy landscape. Critically for this paper, they are also to submit an action plan that includes:

- outline plans to rationalize partnerships;
- a framework for streamlining the delivery of economic development services to remove overlap and duplication, improve effectiveness and release resources for new services;
- the setting of targets for best value in economic development services;
- a clear statement of the respective roles and responsibilities of partner agencies; and
- a monitoring and evaluation framework that is aligned with the 'community planning framework.' (Lloyd and Illseley, 2001).

Much of the activity of the LEFs so far has been conducted directly between partner organizations locally, in most cases without wider involvement or dissemination of plans. Undoubtedly the restructuring and reconfiguration of a number of government agencies to create non-departmental public bodies in, for example, careers guidance and advice (*Careers Scotland*), in housing and social inclusion (*Communities Scotland)* and to deliver improved labour market information (*Future Skills Scotland*) has diverted resources and attention away from the LEFs. However, although the establishment of those new national networked bodies has been a priority, there is still commitment to local partnership working through the embryonic LEFs.

To summarize: in recent times Scotland has been applying an economic development strategy informed by endogenous growth theory but the crowded institutional landscape that has developed to deliver this strategy demanded reform. The Government has proposed a local approach to arranging the delivery of business development services. The elements of this framework: the adoption of the regional development agency model, the promotion of new SMEs and the encouragement of partnerships between agencies, are key characteristics of the Scottish approach which can also be identified in many other parts of Europe, even, as it turns out in Poland, the focus of our attention below.

The SME Sector in Poland

We begin by examining economic strategy, institutional frameworks and support measures for SMEs at both national and regional levels in Poland. At the regional level our focus, as elsewhere in this volume, will be on Upper Silesia (*Śląskie voivodship*). We then contrast the Polish with the Scottish experience noting the problem of institutional overlap for small business support.

The main aim of the empirical investigation was to undertake a pilot study of selected institutions at both national and regional levels involved in SME policy formulation and support. This pilot project is the first phase of a wider programme of comparative research that includes institutional support for SMEs in Scotland and Central Eastern Europe (mainly Poland). A case study methodology was adopted for this phase to develop and then establish a range of specific research questions for the next stage of the project. Key actors across organizations were interviewed and secondary data gathered from national and regional sources.

The rapid expansion of the SME sector in Poland at the beginning of the 1990s was very much the result of a 'grass roots' and 'bottom-up' process rather than the outcome of national policy. The number of SMEs registered officially increased from 1,980,000 in 1990 to 3,007,400 in 1999 (excluding agriculture, forestry, hunting, fishing and fisheries). The majority were established using personal and family resources. This process led to an unprecedented increase in the number of micro and small enterprises. Enterprises with up to five employees account for 92.8 per cent of all Polish SMEs, small firms (6-50 employees) represent 6.1 per cent, and medium-sized enterprises (51-250 employees) comprise 0.9 per cent of the total (Ministerstwo Gospodarki, 2000, p.16). As in other Central and East European economies, SMEs in Poland predominate in certain sectors such as trade and repairs (38.1 per cent), manufacturing (12.7 per cent), real estate and associated activities (12.2 per cent), and construction (10.9 per cent). In terms of origin, three SME types can be distinguished: start-ups, established businesses (pre-1990s), and spin-offs from large state-owned companies. SMEs mainly operate in metropolitan areas and regions with a well-developed industrial base. Towards the end of the 1990s, one-fifth of all SMEs were located in the three largest urban centres of Łódź, Warsaw and Katowice. Therefore, the *Śląskie voivodship* (which incorporates Katowice) has a strong SME sector. It provides a case study region with diverse institutional structure where lack of co-ordination across organizations is also evident.

National SME Strategy

From the mid-1990s it was clear that the SME sector was crucial to the sustainable economic development of Poland in terms of its contribution to GDP, job creation and the containment of increasing unemployment.[7] The Polish government also realized that the SME sector needed more support at both national and regional levels with appropriate policy and institutional development. At national level, the breakthrough came in 1995 when the government of the day proposed that SME policy should be based on four main instruments: legal, financial, institutional, and those incorporating information and training programmes.[8]

Some of the legal instruments have been enacted and implemented. These concern mainly the tax code, banking and guarantees law and the legal framework for business registration. The tax code proved to be the most controversial and was later amended. Other instruments have only recently come into effect, and their impact on SME development is yet to be assessed. Among the financial instruments, the Loan Guarantee Fund, established in 1995 and subsequently transformed into the National Loan Guarantee Fund, is of major importance. From 1997, 15 local or regional funds were established to provide SME loan guarantees. Nevertheless poor management, limited funds and inadequate information regarding the funds meant that too few SMEs used this facility. The Polish Foundation for the Promotion of SMEs (PFPSME) was the most significant institutional initiative. Its main remit was to co-ordinate activities targeting domestic and foreign aid programmes in support of SMEs. It established the *Krajowy System Usług (National Services System)*, a network of SME support institutions throughout the country. The foundation's initiatives have been absorbed by its successor, the Polish Agency for Enterprise Development (PAED).

Measures to support knowledge and learning for SMEs are implemented by a range of state institutions including Centres for Business Promotion, Regional Development Agencies, Regional Development Foundations and a Business Information Network that operates mainly at regional and local levels. However, the most comprehensive range of services is provided by organizations that belong to the *Krajowy System Usług*.

All of this helped to improve SME support but some major barriers still blocked the sector's development.

In 1999, another Polish government formulated a new SME policy. First, a central strategic stance towards SMEs was adopted. This extended support for

---

[7] Employment in the SME sector increased by 19 per cent over 1993-99 and the SME contribution to Poland's GDP grew steadily from 44 per cent in 1996 to 51.5 per cent in 1999. By 1999, SMEs provided 63.8 per cent of all non-agricultural employment. (Ministerstwo Gospodarki, 1999, pp. 5 and 17).

[8] The policy was set out in *Małe i średnie przedsiębiorstwa w gospodarce narodowej*.

small and medium size businesses as an integral part of labour market, regional and sectoral policies. The new policy thrust addressed the main areas thought to have been detrimental to SME development in the past: fiscal policy, financial regulations, insurance and banking.[9] The proposed strategy was then sub-divided into three intermediate objectives: increasing competitiveness, stimulating exports, and raising investment capabilities (Ministerstwo Gospodarki,1999, p.9). However the proposals only partially addressed SME needs. They proved to be effective only in selected industries and regions. It also became clear that successful implementation of those strategies required co-operation between governmental institutions, business communities and local self-government. This process of what seems to be movement towards a partnership-type model may suggest that the Scottish experience has some utility for Poland and its regions. Second, at the end of the 1990s, the government's SME policy began to be strongly influenced by EU accession resulting in the need for better co-ordination between ministries and other institutions with small business responsibilities.

At regional level, since 1999, self-governing bodies in the new voivodships have assumed responsibility for economic development and regeneration strategies. Support for SMEs is regarded as an important element of such programmes, especially in those regions that are restructuring. In Upper Silesia two programmes – Coal Mining and Iron and Steel Restructuring – were specifically designed, with SME initiatives, to ease pressure on the local labour market resulting from extensive heavy industry redundancies.

However, in the 1990s, central government policies with a bearing on SMEs were in fact not directly aimed at the SMEs themselves. The three main areas of government strategy – labour market, rural and regional policy – incorporated SMEs only indirectly. (*Polska Fundacja* ... 2000, p. 140).

Institutional Support at National Level in 2000

Figure 10.2 illustrates the range of institutions involved in SME support. However, it does not articulate the relationships between them as they are not well defined. Some institutions have a close working relationship. These are likely to be most apparent between the enterprise development agency (PAED), the parliamentary committee for SMEs, and the Ministry of the Economy; especially since the latter has recently assumed a more active role in SME policy in line with Poland's accelerated efforts on EU accession. PAED is very much a 'hands on' organization directly involved in the implementation of projects addressed directly to SMEs. From the policy viewpoint the most important institutions are the Ministry of Economy and PAED. However, as Poland approaches EU accession institutions need to co-ordinate their activities to ensure that the SME sector meets the accession criteria and becomes more competitive internationally.

---

[9] This policy was set out in *Kierunki działań Rządu wobec małych i średnich przedsiębiorstw do 2002 roku.*

Figure 10.2   National Support for SMEs in Poland

As the organizational framework for SMEs evolves, further institutions are being established at national, regional, county, and municipality levels. According to the PAED's database (*Polska Fundacja* ...2000, p.205) some 1,508 organizations supported SMEs throughout the country in 1999 (this is an underestimate since not all support organizations are registered with PAED).

The PAED registered institutions can be divided into four major groups. Table 10.1 indicates that there are two main categories of organization. First, the voluntary, self-governing, and non-regulated range of institutions comprising 60 per cent of the units offering SME support provision; and second, state sponsored organizations for SME services and links to the education sector, 40 per cent of the total number. Substantial funds for the SME sector have been received from various bi-lateral and multi-lateral aid programmes (for example, EU, USAID, and the British Know How Fund). The contribution of organizations set up with foreign (and particularly EU) assistance cannot be overestimated as they provide expertise that could not otherwise be found in Poland.

Poland's continuous efforts aimed at EU accession and compliance have led to a proliferation of institutions and organizations delivering business support services and have encouraged some co-operation between agencies at national and regional levels in the SME sector. Such institutions are key components of the attempt to create and disseminate knowledge. However, there is a real danger of duplication of activities due to still limited co-ordination as well as a concern about the capacity of regions and firms to absorb this devolved system. Recent developments in Scotland described earlier in this chapter may alert the Polish policy-makers to impending difficulties and serve as useful examples of how to overcome these problems.

Table 10.1   Spread of Institutions Supporting SMEs in Poland (%)

Type of organization	%
Organizations of entrepreneurs	37.5
Non-governmental institutions	23.4
Institutions providing services for SMEs	23.1
Research institutes and academic institutions	15.9
Total (1,508 units)	100.0

Source: Raport o stanie małych i średnich przedsiębiorstw w Polsce w latach 1998-1999, p. 205.

SME Development in the Śląskie Voivodship

Shifting attention from the national to the regional scene we focus on Śląskie and examine the strategy, institutions and measures supporting the region's SME sector. The *Górnoślaska Agencja Rozwoju Regionalnego* (GARR – Upper Silesian Agency for Regional Development) is, as we will see, an institution that plays a key role in regional SME support.

The SME sector in the old Śląskie voivodship (that is, the administrative structure pre-1999) had some unique features. Due to its location close to the Polish/Czech and Polish/Slovak borders, the 'grey economy' flourished throughout the 1990s by exploiting price differentials between the countries. At an early stage of transition this phenomenon played an important role in both job and GDP creation (Grabowski, 1995).[10]   Small and medium sized companies grew rapidly providing goods for sale from outlets located close to the border. More recently, it was hoped that the expansion of SMEs in the region would be fuelled by former miners and steel workers investing their redundancy pay in local business ventures. However, this did not happen for a variety of reasons. First, the mentality of former miners is very special, perhaps elitist: they are said to belong to a privileged group in Poland. Second, they benefited from the protection of strong trade unions and they are used to working in large enterprises. Third, it has been difficult to attract former miners (and even more difficult with steel workers) to some of the SME support schemes (loans, advisory services and so forth).

Regional SME Strategy

Figure 10.3 shows those organizations supporting the region's SMEs. Three broad groups exist. First, those that are based in self-government (Marshall's) and central government (Voivod's) offices.   Second, governmental sponsored

---

[10] Grabowski, Z. (1995), estimated that the contribution of the grey economy to Poland's GDP could be close to ten per cent.

organizations incorporating foreign aid, including the Upper Silesian Development Agency (GARR) and the Upper Silesian Agency for Enterprise Privatization (GAPP). Third, a group of institutions with more diverse origins such as chambers of commerce, academic research institutions and the Upper Silesian Fund, a financial institution established in 1995 in the Regional Contract for Katowice with the State Treasury as one of its shareholders.

In Silesia (as in other regions, see Gorzelak and Jałowiecki, 2000) the Marshall's office plays the most important role in shaping the region's future. The Marshall's office, in co-operation with local government, trade unions and other regional interests groups and stakeholders prepares the strategy document for the voivodship. This is an inventory of achievements and problems facing the region as well as a thorough analysis of its current socio-economic development. The region's development path for the next decade is mapped out. Priorities are established. One priority, with an SME focus, aims at 'an increased innovative potential and competitiveness of the economy, including small and medium size enterprises' (*Sejmik Śląskie*, 2000 p. 128). Within this priority five 'action targets' specific to SMEs are outlined: a further development of the craft sector, general SME support and promotion and improvement in its infrastructure, further assistance leading to more SMEs gaining patents and licenses, and support in modern management techniques, marketing and finance.

The Strategy recognizes that policy delivery is better channelled through knowledge acquisition measures. An increase in entrepreneurial activity and a well established SME sector are viewed as vital for regional success, and as partially offsetting the negative effects of the region's regeneration and transformation from a mono-cultural and mono-functional development path to one based on a more modern economic structure (Fedorowicz, 2000).

At the local level, an important institutional arrangement for SME support is the network of 17 Local Agencies (Local Segments) resulting from the Regional Contract for Katowice. The Regional Contract, pioneered in 1995, was a novel approach to regional development, one of the first (if not the first) involving the concerted activity of central government, local government, trade unions and other groups with regional interests and a stake in regional development. 'Local Segments' are unique to this region and were created in order to assist the transformation process (mainly coal and steel restructuring). They have three main aims: to help combat regional unemployment through job creation and retraining in conjunction with the local labour offices; to assist the development of entrepreneurship and to support redundancy negotiations.

The GARR development agency, the most dynamic and important among the newly created institutions in Upper Silesia, works in partnership with a large number of institutions at both local and regional levels, and played a key role in the preparation of the *Strategy for Śląskie Voivodship for 1998-2002*. GARR's activities focus mainly on co-ordination rather than control of local and regional economic development.

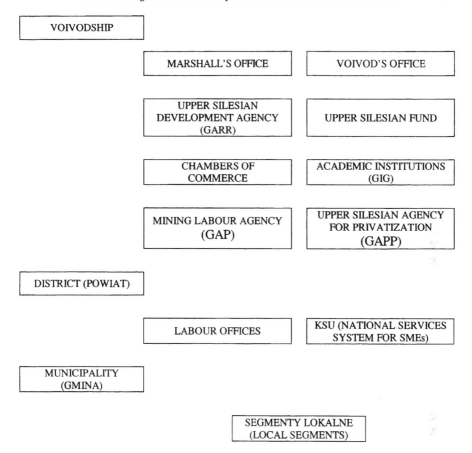

Figure 10.3 SME Support in Silesia

Support Measures for SMEs[11]

GARR is also strongly involved in SME support (especially financially) and administers the following schemes:

- a loan fund to assist job creation in sectors designated in the 'Regional Contract'. Some 75 per cent of this fund's financial support comes from central government and the rest from GARR. The loan fund helped create 222 new jobs in municipalities suffering from high structural unemployment in 1999;
- a loan fund for economic activities based on the *Moje małe przedsiębiorstwo'* (My small business) scheme. Unlike the first fund, this is entirely a GARR scheme,

---

[11]This section is based on an interview with GARR's reperesentatives (Mrs Iwona Czaplikowska and Mr Grzergorz Giemza) in February 2002.

providing financial support to local businesses that are winners of the *Moje małe przedsiębiorstwo* competition;

- loans to supplement SME's own investment. Again this was a unique solution introduced and financed from GARR's own funds and has proved very popular as it offered free advice and an interest rate much lower than from commercial banks;
- the *Inicjatywa* (Initiative) was part of the Phare 1998 Programme and had two main components. First, a loan fund to finance ventures aimed at the creation of sustainable new work places for former miners. Loans at preferential interest rates were allocated for job creation for unemployed miners, or for businesses start ups providing self-employment for them or their spouses. Some 345 loans were approved for 322 borrowers (130 for SMEs and 215 directly to former coal miners or their spouses). The number of loans exceeded borrowers as some obtained several loans. The scheme provided employment for 530 former coal sector employees in Silesia. A second scheme offered former miners, as well as for existing SMEs, co-financing of advisory services in business start ups creating new jobs for miners. A total of 428 advisory services were provided, including 279 for miners or their spouses and 149 for SMEs.

Inicjatywa proved to be successful for two reasons. First, the combination of loan and advisory services located in one institution (GARR) for the coal sector was a successful formula. Second, a strong point of the programme was a very efficient advertizing campaign initiated by GARR at the start of the project in January/February 2000. However, 'Initiative' also had some failings (mainly its loan component). A more stringent screening process of potential borrowers needs to be applied in the future. There were cases where individuals could have borrowed money from the banks but it was cheaper for them to use the loan scheme. Monitoring also needs to be tightened up. In some cases checking whether SMEs followed the conditions for recruitment of former coal miners was not applied.

Silesian institutions have also gained much experience through administering PHARE funds. It is clear that Silesia is well served by a range of institutions and financial schemes dedicated to providing SME support. They have diverse origins, sometimes local bottom-up initiative, sometimes government support or backed by foreign aid. But the region's SME institutional framework and loan provisions are also a response to the unique local need to combat the negative impact of coal and steel restructuring.

Institutions Supporting SMEs

The devolution of state and public administration with the creation of 16 new regions in January 1999 is highly significant for the new *Ślaskie voivodship*. Its territorial composition was dramatically reshaped by bringing together three large, formerly separate, voivodships – Katowickie, Bielskie and Częstochowskie – with profound repercussions for institutional support for regional economic development. First, a number of institutions were created as part of the 'Regional Contract', providing an institutional basis crucial for the region's development. In Żory, the *Agencja Rozwoju Przedsiębiorczości SA* helps to alleviate the negative consequences of coal mine closures. Bytom, with its loss making coalmines, is supported by the *Agencja Inicjatyw Lokalnych SA* that assists with implementing a programme of spatial development based on the use of IT (*Rzeczpospolita*, 1998). Similar organizations were established in Raciborz and Ruda Ślaska with the common aim of supporting local entrepreneurship. Second, the January 1999 administrative reform brought together all the institutions from the previously disparate voivodships of Bielsko, Czestochowa and Katowice into one regional structure (*Sejmik Województwa Śląskiego*, 2000, p. 118). Third, various bilateral aid programmes (for example UK, US, and Swiss) and consecutive EU sponsored schemes were instrumental in setting up institutions aimed at administering funds and advice to SMEs. Thus, with its large number of institutions for SME support the Śląskie voivodship provides ample material for an in depth case study, with the possibility that the problem of overload, duplication and such like, as found in Scotland, will also be a feature of the local scene.

Conclusion

The process of economic restructuring and arresting regional decline has been underway in the European Union for some time. It is generally thought that a coherent economic development strategy has two critical elements. First, it needs a devolved and accountable arrangement for policy delivery. This requires a local, usually regional, administrative structure supported from the central budget supplemented by local revenue raising powers. Second, it needs effective co-ordination, drawing together multi-agency activities to make optimal use of the resources devoted to economic and enterprise development.

For Central and Eastern Europe, and Poland in particular, lessons may be learned from the Scottish case where congestion of services had to be rationalized, and agency partnerships became a key element in delivering regional economic development policy. Those aspects of policy implementation in Scotland may have specific implications for Poland. First, to capitalize on potential cluster effects where a more effective regional effort for SME support can be applied. Second, encouragement of a coherent and cooperative strategy between a vast array of agencies is desirable. Third, the introduction of a rigorous range of criteria for establishing the effectiveness and 'value for money' derived from each policy that supports or promotes indigenous enterprise is needed. Fourth, the key to sustained endogenous economic growth is the creation and dissemination of knowledge, and this must be part of any SME or regional strategy.

References

Amin, A. and Thrift, N. (1994), 'Living in the global', in A. Amin and N. Thrift (eds) *Globalization, Institutions, and Regional Development in Europe,* Oxford, Oxford University Press.

Armstrong, H. (1997), 'Regional-level jurisdictions and economic regeneration initiatives' in M. Danson, G. Lloyd, and S. Hill (eds), *Regional Governance and Economic Development,* London, Pion.

Boltho, A. Holtham, G. (1992), 'The Assessment: New Approaches to Economic Growth', *Oxford Review of Economic Policy 8,* 4.

Cameron, G. and Danson, M, (2000), 'The European Partnership Model and the Changing Role of Regional Development Agencies. A Regional Development and Organization Perspective' in Danson, M. Halkier, H. and Cameron, G, *Governance, Institutional Change and Regional Development,* Aldershot, Ashgate.

Danson, M. Halkier, H. and Damborg, C. (1998), 'Regional development agencies in Europe; an introduction and framework for analysis' in *Regional Development Agencies in Europe,* London, Jessica Kingsley Publishers.

Danson, M. Fairley, J. Lloyd, G. and Turok, I. (1999), 'The European Structural Fund Partnerships in Scotland - New Forms of Governance for Regional Development?', *Scottish Affairs.*

Fedorowicz, R. (2000), 'Ambitna strategia, Województwo jutra' (An ambitious strategy for voivodship of tomorrow), *Gospodarka Śląska,* November.

Freeman, C. (1994), 'The Economics of Technical Change', *Cambridge Journal of Economics, 5.*

Gorzelak, G. and Jałowiecki, B. (2000), 'Metodologiczne podstawy strategii rozwoju regionu na przykładzie województwa lubuskiego' (The methodological basis of regional development strategy, the Lubuskie example) in *Studia Regionalne i Lokalne,* No.3.

Grabowski, Z. (1995), 'Szara strefa a transformacji gospodarki' (The grey economy and transformation), Instytut Badań nad Gospodarką Rynkową, *Transformacja Gospodarki,* No.58, Gdansk.

Grabher, G. and Stark, D. (1997), 'Organizing Diversity: Evolutionary Theory, Network Analysis and Postsocialism', *Regional Studies 31,* 5.

Gregerson, B. and Johnson, B. (1997), 'Learning Economies, Innovation Systems and European Integration', *Regional Studies 31,* 5.

Halkier, H. and Danson, M. (1997), 'Regional development agencies in Western Europe: a survey of key characteristics and trends', *European Urban and Regional Studies 4,* 3.

Heinelt, H. and Smith, R. (eds), (1996), *Policy Networks and European Structural Funds.* Aldershot, Ashgate.

Hooghe, L. (ed), (1996), *Cohesion Policy and European Integration. Building Multi-level Governance,* Oxford, OUP.

Hodgson, G. (1988), *Economics and Institutions.* Polity, Cambridge.

Kafkalas G. and Thoidou E. (2000), 'Cohesion Policy and the Role of RDAs in the Making of an Intelligent Region. Lessons from the Southern European Periphery', in Danson, M. Halkier, H. and Cameron, G.

Lloyd, M.G. and Illsley, B.M. (2001), 'Community planning in Scotland: Prospects and potential for local governance?' in *Divided Scotland,* M. Danson, J. McCarthy and D. Newlands (eds), Aldershot, Ashgate, forthcoming.

Lundvall, B.A. and Johnson, B. (1994), 'The Learning Economy', *Journal of Industry Studies 1,* 2.

Ministerstwo Gospodarki i Handlu, (1995), *Małe i średnie przedsiębiorstwa w gospodarce narodowej* (SMEs in the national economy), Warszawa.

Ministerstwo Gospodarki, (2000), *O stanie realizacji, 'Kierunków działań Rządu wobec małych i średnich przedsiębiorstw do 2002 roku'*, Warszawa, listopad.

Ministerstwo Gospodarki, (1999), *Kierunki działań Rządu wobec małych i średnich przedsiębiorstw do 2002 roku* (Government Policy Guidelines for Small and Medium-Sized Enterprises until 2002), Document approved by the Council of Ministers on May 11.

Morgan, K. (1997), 'The Learning Region: Institutions, Innovation and Regional Renewal', *Regional Studies 31*, 5.

Newlands, D. and McCarthy J. (1999), *Governing Scotland: Problems and Prospects. The Economic Impact of the Scottish Parliament*, Aldershot, Ashgate.

Van der Ploeg, F. Tang, P. (1992), 'The Macroeconomics of Growth: An International Perspective', *Oxford Review of Economic Policy 8*, 4.

Polska Fundacja Promocji i Rozwoju Małych i Średnich Przedsiębiorstw, Warszawa, (2000), *Raport o stanie małych i średnich przedsiębiorstw w Polsce w latach 1998-1999* (Report on Polish SMEs 1998-1999).

Roberts, P. (1997), 'Sustainability and spatial competence: an examination of the evolution, ephemeral nature, and possible future of regional planning in Britain', in Danson, M. Hill, S. and Lloyd, G. (eds), *Regional Governance and Economic Development*, European Research in Regional Science 7, London, Pion.

Rzeczpospolita, (1998), 'Nie trzeba startować od zera' (There's no need to start from zero), September 24.

Scott, M. (1992), 'A New Theory of Endogeneous Economic Growth', *Oxford Review of Economic Policy 8*, 4.

Scottish Enterprise (2000), *The Scottish Business Birth Rate 2000, Improving the Business Birth Rate, Scottish Enterprise 2000*.

Scottish Executive, (2000), 'Local Economic Forums: Draft National Guidelines - Consultation Draft', Enterprise & Lifelong Learning Department.

Sejmik Województwa Śląskiego, (2000), *Strategia rozwojowa województwa Śląskiego na lata 2000-2015*, (Development Strategy of the Śląskie voivodship), Katowice.

Steiner, M. (1984), *Alte Industriegebiete - theoretische Ansätze und wirtschaftspolitische Folgerungen*, Dissertation, Uni. Graz.

Storper, M. and Scott, A. J. (1995), 'The Wealth of Regions', *Futures, 27*.

Storper, M. (1992), 'The Limits to Globalization: Technology Districts and International Trade', *Economic Geography 68*.

Storper, M. (1995), 'The Resurgence of Regional Economies, Ten Years Later: The Region as a Nexus of Untraded Interdependencies', *European Urban and Regional Studies 2*.

Wannop, U. (1995), *The Regional Imperative. Regional Planning and Governance in Britain, Europe and the United States*, London, Jessica Kingsley Publishers.

Whittam, G. and Kirk, K. (1996), 'The Business Rate, Real Services and Networking: Strategic Options', in Danson, M. *Small Firm Formation and Regional Economic Development*, Routledge, London.

Chapter 11

Enterprise Development in Upper Silesia – A Difficult Path in Regional Restructuring

Bogumił Szczupak

Introduction

Economic restructuring in Silesia is taking place alongside the transition from a centralized to a market economy, a shift from an economy based on transformation of materials, where huge industrial sectors (coal-mining, power, metallurgy, arms) dominated, towards an economy increasingly utilizing regional non-material factors and technologies brought in through FDI. All this occurs as the regional economy is forced, through globalization, into greater openness. In addition, co-operation between firms in the region collapsed. This was particularly so in the supply of investment goods and intermediate inputs to mining and metallurgy but it also affected relations between science, innovation and research centres and manufacture and services. At the same time new democratic institutional and organizational structures are being developed at the level of *gminas, powiats* and the voivodship all of which initially seemed unwilling to support and develop the small and medium sized enterprise sector. As a regional structural policy is shaped it is important to distinguish macro- from microeconomic conditions, where the main elements are privatization and the restoration of a culture of entrepreneurship. There is a need too for complementarity between policies for sectoral restructuring and the policy of regional and local re-conversion (Quévit, 1995).

Regional Restructuring and Globalization

The processes of regional restructuring is leading to a concentration of the activities of global sector enterprises in the most attractive towns leading to increasing polarization across urban centres as they become divided into those that are developing and those that are stagnating. Restructuring is concentrated in the towns of the Katowice agglomeration. At the same time the number of people working in the public and private sector is equalizing almost everywhere. Private sector employment predominates only in four towns (Katowice, Siemianowice, Sosnowiec and Tychy) and in the case of Tychy it is 2.5 times greater than in public sector.

Privatization and direct foreign investment appear to have changed the Katowice agglomeration to one where two groups of towns can be distinguished. The first group is characterized by relatively dynamic development and includes Katowice, Tychy, Bieruń, Lędziny. They have over 40 per cent of private sector enterprise assets within the Katowice agglomeration and 21 per cent of the population. The second group includes Gliwice and Zabrze with 19 per cent of enterprises assets and nearly 17 per cent of population. These two groups of towns, the inner centre of the agglomeration, succeed because they provide mobile, global companies with financial and administrative advantages such as tax reliefs and good infrastructure. The situation of towns dominated by power and steel is clearly more difficult. At present, regional authorities are in a weak position, with few effective instruments to influence the investment decisions of firms. There is no organized regional approach to develop post-industrial areas, revitalize town centres or to counteract the degradation of the infrastructure.

Enterprise Development – Progress and Barriers

The main barrier to supporting development of native small companies in Silesia is the poor quality of entrepreneurship and the slow process of the regeneration of any enterprise culture among those employed in large industrial enterprises. Polish local authorities pay too little attention to these matters. The *gmina* (municipality) administers and organizes rather than manages development (Wagner, 1997). Over 308,000 people run their own businesses. However, activities are guided almost exclusively towards the local market, marked by low technology and little capital outlay. Almost 39 per cent of businesses are in trade and repair. Some 11.6 per cent are in construction. Very few create employment. Only 62,000 people became employers. The rate of business start up has also clearly declined. Entrepreneurship encounters serious development barriers. Outer barriers include the deterioration in the general economic climate, the too tough competition from outside investors and limited possibilities of gaining financial support. The inner barriers include poor ability, a low level of professional competence and lack of those essential resources needed to start-up.

Entrepreneurial behaviour across local communities is also diverse. Some towns have relatively strong entrepreneurship (perhaps based on a historical tradition of stronger merchant traditions). They include Bielsko-Biała, Cieszyn, Żywiec, Rybnik, Racibórz, Gliwice. But elsewhere the picture is far from rosy. There is little entrepreneurial activity among the inhabitants of those towns that became industrial during the 19th century, with little in the way of a merchant tradition, and where educational attainments are low.

SME development in Silesia also meets a range of fairly typical barriers. They include limited access to capital, to information on standards of products and technologies. It is not easy to establish co-operation links with larger, especially foreign companies. They prefer to co-operate with companies from the home country. There is insufficient economic, organizational, and marketing knowledge and poor access to support systems organized by regional and local authorities.

The process of organization of SME supporting institutions has ceased in Silesia. The main reason is the lack of legal solutions to regulate public and private partnership. So far, institutions supporting SMEs have had a commercial nature or have been tightly centralized and dependent on subsidies from central budget. Legal constraints on economic activity of *gminas, powiats* and voivodships effectively obstruct the emergence of good conditions for an entrepreneurial environment at the local level. The existing institutional infrastructure which should stimulate and guide enterprise has effectively exhausted its possibilities to do so.

Shaping Entrepreneurship – A Strategic Challenge for Regional Authorities

An enterprising environment is shaped through partnership of those institutions involved in local development (Szczupak and Czornik, 2001). The logic of partnership traces a learning path through the local environment that adds dynamism to the mutual expectations of local actors. Partnership can also be viewed as a form of co-governing (Danson, Halkier and Cameron, 2000). In the reality of towns in Silesia local authorities too rarely initiate partnerships. The authorities are too slow in 'de-municipalizing' local resources through partnerships and so limit the tapping of local development potential. Shaping networks of co-operation yields competitive advantage through disclosure of complementary investment and product development activity. It also reduces transactions costs. It may lead to common initiatives in various activities. From the viewpoint of local competitiveness such co-operation results in an improved investment climate. In the reality of Silesian towns, one observes great diversity in entrepreneurship, partly the result of underdeveloped partnerships.

Supporting the potential of existing companies to grow requires in particular the support of innovation. Entrepreneurs can learn quickly about new technologies and new ways of applying knowledge through sharing mutual experiences and information. Fostering such an innovative entrepreneurial environment promotes local (town) competitiveness, encourages small firms to grow, is good for exports and generates new places of work. From the point of view of creation of town competitiveness institutional frameworks for wider partnerships are needed with more civic agencies, social organizations and others through which dialogue will multiply the possibilities facing private entrepreneurs. Partnership institutions often take responsibility, as in Scotland, for skills improvement, stimulating entrepreneurial attitudes and behaviour in the local community. In the reality of contemporary Silesia local authorities too rarely initiate partnership or indeed make any civic initiatives. The authorities fail to make use of local resources through partnership and thus, limit local development potential.

References

Quévit, M. (1995), Polityka regionalnej restrukturalizacji w Krajach Europy Środkowej i Wschodniej: wnioski z polityki realizowanej przez kraje OCDE (Regional restructuring policy in Central and Eastern European countries: conclusions from policy implemented by OECD countries) in *Studia Regionalne* [Regional Studies], Academy of Economics, Poznań.

Szczupak, B. and Czornik, M. (2001), *Środowisko przedsiębiorczości a konkurencyjność miasta.* (Entrepreneurship environment and town's competitiveness), Conference resources. Academy of Economics, Katowice - Ustroń.

Danson, M. Halkier, H. Cameron, G. (2000) *Governance, Institutional change and regional development*, Ashgate, Aldershot.

Wagner, P. (1997), 'Kilka uwag na temat ustawy o gospodarce komunalnej', (Some comments on the bill on the municipal economy), mimeo.

Chapter 12

Enterprise Development: Lessons for Enterprise and Enterprise Clusters from Scotland's Experience of Regional Policy

Mike Danson and Geoff Whittam

Introduction

Scottish Enterprise (SE), the development agency for Scotland, launched a 'Clusters Approach' for its key industrial sectors in 1998 (*Scottish Enterprise* 1998). In the document it argued that if regions and nations are to be competitive in the new century they must be at the forefront of knowledge production. SE readily acknowledges that to compete the information and ideas created and developed through clustering depend on strong partnerships. The knowledge-based economy and its synergies hinge on 'working to build trust and a shared vision' (*Scottish Enterprise*, 1998, p. 1). SE envisages strong partnerships of customers, suppliers, competitors, universities, colleges, research bodies and the utilities, in other words a private/public sector mix. This is hardly surprising given the nature of the output (information) being produced, and the private sector's well-known tendency to under-invest in research and development. This under-investment occurs due to the large fixed costs associated with the development of a new product or process, the inability of the innovator to reap all the innovation benefits and the uncertainty as to whether a new product or innovation will be successful. Given that knowledge production creates spillovers, that much of the knowledge produced is tacit, and that new products and processes will benefit from critical commentary from all the players involved in the partnership, there is a strong spatial element in winning competitive advantage.[1] The goal is to establish a 'learning economy' within the globalized economy. The spatial dimension may be at a regional or a national level or indeed can transcend state/regional boundaries. Much of this is readily accepted in the existing literature.

The Scottish economy already has examples of partnership frameworks that could be adopted to assist the success of SE's clustering strategy. The most successful is the Strathclyde European Partnership (SEP). The SEP has received widespread acclaim

---

[1]Tacit knowledge is knowledge acquired by experience that cannot be codified, that is, easily transferred. To acquire tacit knowledge it is neccessary to be involved in the production of that knowledge.

(Danson, Fairley, Lloyd and Turok, 1997) as a model for bringing together the key players – both public and private – in promoting regional economic development.

In this chapter we examine evolving partnership governance structures in Scotland at a regional and county level highlighting both good and bad practice. Before that however we begin by placing the evolving clustering strategy in the context of Scotland's economic development. We conclude with policy implications and warning signs for the development of a clustering strategy based on the learning economy in Scotland.

Scotland – On the European Periphery

While the period since the UK joined the EU has refocused trade, industry and the regions of the British Isles in many ways it also exacerbated a much longer decline in Scotland, Wales and the north of England. Indeed, over 'most of this century [the twentieth century] Scotland has been declining relative to the rest of the United Kingdom and, by extension, the rest of Europe' (Danson 1991, p. 89). As one of the first industrialized regions of Europe and so of the world, Scotland experienced both an early period of growth and a long history of stagnation. Its growth and development in that earlier period was built on strong networks as a leading economic historian (Slaven, 1975, p. 182) puts it,

> A community of interests was growing among steelmakers and shipbuilders, and this was a link more strongly developed at a later date ... The demands of the shipyards boosted the growth of steel and dictated the changing production patterns of the pig-iron and malleable-iron producers. Marine engineering and shipbuilding lay at the centre of a complex concentration of heavy engineering and finishing trades.

Behind the development of the Scottish economy, and of Clydeside in particular, were factors associated with the British Empire. Britain's role as an imperial power, based on naval supremacy, prompted the establishment of the coal and steel industries of central Scotland. The regional economies and, through linkages and migration, the rest of the nation, became inextricably dependent on the trading and military position of the UK as a whole. Clydeside experienced very significant growth in industry and population in the years up to 1914. Records on shipbuilding tonnage (Slaven, 1975) confirm that the region was at the heart of the Empire in terms of industrial output and importance. The First World War was perhaps the watershed although the seeds of external destruction may have been present before then: the decline of British power and imperial markets exposed a rapid and deep structural imbalance in the Scottish economy. Massive unemployment, poverty, deprivation and emigration marked the period up to the next World War, with Glasgow suffering some of the worst urban slums in history (Damer, 1990). Since 1930 this industrial legacy ensured that Scotland was subject continuously to a broad set of economic policies targeted at relieving the worst effects of the rundown and closure of the staple industries – steel, coal, shipbuilding, heavy engineering and textiles.

Throughout the twentieth century, Scotland – and the rest of the periphery of Britain – was in almost constant receipt of various forms of regional development aid. In more recent times regional economic policies, based on restructuring the old industrial areas, and the decongestion of overcrowded housing and manufacturing areas, have become much more concerned with encouraging foreign direct investment. The state has increasingly withdrawn from direct involvement in production as nationalization has yielded to privatization. Meanwhile, through merger and take-over, the degree of external control and ownership of Scottish industries has steadily increased. Differential rates of decline and growth of native and foreign companies has also been an important factor. Research suggests that such changes, in complex ways, may slow rates of new firm formation and indigenous development (Ashcroft, Love and Schouller 1987). Moreover, as time has passed, output, trade and investment in Scotland have become more narrowly dependent on a few key sectors.

The regional policy focus on inward investment alongside low rates of endogenous growth has created in Scotland a 'branch plant economy' in place of the former heavy industrial clusters. With few local supplier or purchaser linkages, and an absence of many of the higher level business functions such as R&D, finance, marketing and corporate strategy, branch plants create local jobs but little else. In particular, there are minimal relations established or developed between inward investors and the local SME sector.

In recent years electronics (especially computers) and whisky have accounted for over half of Scotland's exports (Scottish Council Development and Industry, 2001), and about 50 per cent of all manufacturing investment. Both sectors are dominated by overseas companies and over 90 per cent of output comes from non-Scottish firms. Over a quarter of manufacturing employment in Scotland is in overseas owned plants, and much of the rest is controlled by UK corporations with headquarters in the south east of England. This degree of domination is often blamed, in a simple way, for the massive restructuring of the Scottish economy since 1979.

Since 1980, employment in Scottish manufacturing has been halved and by 2010 only 11 per cent of all jobs in Scotland are expected to be in manufacturing (Institute for Employment Research, 2001). In key traditional sectors such as coal, steel and engineering the decline in employment was heavier, with some compensation in electronic and electrical engineering. In many sectors: agriculture, fishing, energy, water, manufacturing and construction, employment is historically at its lowest levels in Scotland, with jobs for both men and women disappearing. Only services show any growth over time, and then only in part-time work.

In such narrow economies, dominated by historic legacy, new firm formation and small and medium enterprises have become viewed as a panacea in regional economic development and regeneration programmes, as the mission statements of many agencies, and of the EU in particular, testify. The business birth rate has undoubtedly increased since 1980 (*Scottish Enterprise* 1996) but the recent review of SE's *Business Birth Rate Strategy* (*Scottish Enterprise* 2000) has indicated that this optimistic picture has not been maintained.[2] Questions have also been raised over the ability of companies dependent on the local market to reverse, on their own, long term regional decline. Malecki and Nijkamp (1988), for example, have suggested that uneven development is endemic, with no possibilities to overcome metropolitan core bias through compensation of the periphery, although Vaessen and Keeble (1995) appear to see enough examples of the counterfactual successful SME in the periphery to argue that regional divergence need not be inevitable. In practice, efforts to replace traditional major manufacturing employers have met with limited success and more prosperous areas of the UK have tended to maintain higher rates of business start-up (Ashcroft and Love 1996). Problem regions also have unemployment rates persistently significantly higher than the UK average (Beatty and Fothergill 1999). The nature and form of networking and of the wider business environment is also of critical relevance in determining such comparative performances in regeneration. A key question for the success of the new firm and of the inward investment strategies, separately and collectively, is whether they are conducive to the creation of networking and clusters. Before we examine the implementation of the clustering strategy within Scotland we review the theoretical underpinnings of the policy and its roots in endogenous growth theory.

Theoretical Underpinnings

At the UK level the White Paper, *Our Competitive Future* (DTI, 1998) and the accompanying *Analysis* had as its 'big idea' the importance of knowledge and knowledge creation:

> knowledge has always been important, ... four mutually reinforcing processes are increasing its importance for prosperity ... Information and Communications Technology (ICT) ... increased speed of scientific and technological advance ... global competition .... changing demand. (Analysis 1988, p. 3).

Significantly the *Analysis* distinguishes between codified and tacit knowledge and much of the White Paper is devoted to explaining how the role of institutions, organizations and individuals needs to evolve to capture and develop knowledge, particularly of a tacit kind, to create a competitive successful economy. The *Analysis* underpinning the emphasis on networking and clusters found in the White Paper highlights the findings from endogenous growth theory (*Analysis* 1998 p. 17). It is through co-operation between key players that firms, regions and ultimately the UK economy is going to be competitive. In terms of science and technology, for example, the *Analysis* (p. 7) explains:

---

[2] The Business Birth Rate Strategy was launched by SE in 1993. The aim was to encourage more business start-ups and a more positive climate towards enterprise and entrepreneurship.

To gain access to technology they (firms) must network with suppliers, customers, competitors and other users of similar technologies ... Firms are also increasingly developing links with the academic science and engineering base.

Knowledge and the innovation process has for long been identified as a significant factor in endogenous growth and in accounting for the 'Solow residual' (the unexplained factors contributing to growth and attributed to technological change) (Boltho and Holtham, 1992; van der Ploeg and Tang, 1992; Scott 1992). The key component of the innovative process is the creation of knowledge, a point recognized by theorists and practitioners alike, (see, for example, Gregersen and Johnson, 1997; CEC, 1993). This has focused attention on how firms, regions and indeed nations innovate. In researching this phenomenon, regional scientists have utilized the methodology of the evolutionary economists (Gregersen and Johnson 1997; Morgan 1997; Grabher and Stark 1997; Storper and Scott 1995; Storper 1992 and 1995; and Lundvall and Johnson 1994). Many researchers point to the importance of evolving institutional frameworks as making major contribution to the discovery and utilization of knowledge. Institutions in this sense include both formal organizations, such as regional development agencies, and informal bodies sometimes defined as 'social organization which, through the operation of tradition, custom or legal constraint, tends to create durable and routinized patterns of behaviour.' (Hodgson, 1988, p. 10).

The emphasis from the regional science perspective is that due to the nature of knowledge – much of it being tacit, it is not easily transferable – the system or region that creates it will develop the advantages associated with innovation. Furthermore, Storper (1995) highlights the importance of what he describes as 'untraded interdependencies'. For Storper, the interdependency arises from the way technology is developed. Utilising the evolutionary approach he focuses on the path dependency nature of technological development. 'Technologies, for one thing, are subject to a variety of user-producer and user-user interactions.' (Storper, 1995, p. 204). Where clustering occurs because of some commonality of technological development then 'untraded interdependencies' arise, such as common coded language, norms, customs and practices. These common institutions lead to easier communication and facilitate trust and co-operation. Similarly, Freeman comments,

> Firms learn both from their own experience of design, development, production and marketing and from a wide variety of external sources at home and abroad – their customers, their suppliers, their contractors...and from many other organizations – universities, government laboratories and agencies, consultants, licensors, licensees and others. (Freeman, 1994, p. 470).

This list is not unlike those institutions identified by SE as making up a typical cluster: companies, customers, suppliers, utilities, research institutes, education (http://www.scotent.co.uk).

This cluster strategy is integrated into the overall SE approach to economic development through a number of other strategic initiatives. The recent launch of the 'Know-How' strategy in particular claims to address the need to nurture education, training and learning more generally. In establishing both a Minister and a Committee of Enterprise and Lifelong Learning (CE&LL), the Scottish Executive and Scottish Parliament seem to recognize the critical role of endogenous learning based along indigenous lines.

So, beyond this, whilst original ideas may emerge outside a specific locality their implementation, that is, innovation, can be achieved most efficiently within an area with sympathetic institutions, an area where 'untraded interdependencies' have developed. Gregersen and Johnson (1997) make a distinction between 'the production of knowledge and the utilization of knowledge'. What is significant for Gregersen and Johnson is that,

> learning has become increasingly endogenous. Learning processes have been institutionalized and feed-back loops for knowledge accumulation have been built in, so that the economy as a whole is learning by interacting in relation to both production and consumption. When economies learn how to learn the process tends to accelerate. (Gregersen and Johnson, 1997, p. 481).

This would be the ultimate aim of a networking or clustering strategy based on learning. It would embrace the whole learning process from acquiring new knowledge, the processing of new knowledge into innovative methods of production resulting in new learning for all cluster participants, in turn leading to further advances in new knowledge and so on.

Development of Clustering within Scotland

Scotland, some argue, has generated a model approach to regional economic development, with its strategic partnerships of central and local government, regional development agencies, 'Quangos' (Quasi Autonomous Non-Governmental Organizations), and other players in the public, private and voluntary sectors (Danson et al, 1997). Such partnerships have been established over the last twenty years, especially in West Central Scotland, through the progressive development of multi-agency, multi-year, multi-functional initiatives, task forces and area agreements (Randall, 1987; Moore and Booth, 1986). In Scotland, over the years, Labour Party dominated local authorities have worked closely on a common agenda with the more market oriented *Scottish Development Agency* (SDA) (Danson, Fairley, Lloyd and Newlands, 1990). The model is based on co-operation, experience and trust. This has been critical in producing a model of intervention that is now adopted across the EU. Although individual partnerships have been criticized (Boyle, 1989; Collins and Lister, 1996 – and see too the searing criticism of the Ferguslie Park Partnership by Chik Collins in this volume), the durability and wider international acceptance of the partnership idea testifies to its perceived effectiveness. Even SE's currently fashionable cluster strategy owes much to the partnership model shaped by its predecessor organization the SDA.

It seems to be accepted that partnerships are not only a legitimate and effective way of co-ordinating and focusing the resources of a number of agencies on a problem, but also the favoured approach. Diverse local actors such as further education colleges, housing agencies, health boards and others have joined the mainstream economic development bodies in stressing the benefits and the need to work in formal partnerships with each other in meeting economic and social challenges. The clusters strategy (*Scottish Enterprise*, 1998) stresses the need to position Scotland at the high value end of the market with enterprises active in clusters operating within 'a local economic environment geared to innovation, investment and upgrading' (p. 4). The strategy is supposed to be advanced by a flexible public-private partnership including industry, education and government. Partnership are to devise action plans around which all economic players will gather to promote an integrated development policy. Integration will be extended to the critical issues identified by SE. As the agency puts it,

> the need to capitalize better on Scottish scientific and technological expertise; to help companies build up their research, design and development capacities; to stimulate entrepreneurship; to encourage companies' international ambitions and to build a high-performance transport infrastructure (*Scottish Enterprise*, p. 4, 1998).

The strategy is holistic, based on synergies and linkages. Clusters will be inclusive, embracing both indigenous and inward investment. Technology, innovation and sustainable economic growth and development are stressed, suggesting the transmogrification of *Scottish Enterprise* back into the regional development agency it was in the 1970s (*The Scottish Development Agency*) rather than the business agency of the 1980s and beyond (Danson et al, 1990). This approach is replacing the single agency interventions of the past. To illustrate the point we assess progress in one particular area in Scotland – Ayrshire. Despite the progress here it seems to us that *Scottish Enterprise* does not yet recognize the full implications of the idea – *the industrial district* – that underpins contemporary cluster strategy.[3]

Local Economy Regeneration Partnerships

In Ayrshire, through an initiative being replicated across lowland Scotland, public, business and independent bodies have been brought together in a partnership, *Ayrshire Economic Forum*, to prepare an Ayrshire Economic Development Strategy. The key players here are: the Local Enterprise Company (LEC), Scottish Enterprise Ayrshire, (the locally managed government agency established to co-ordinate the delivery of training and business development services and infrastructure), the three local authorities (East, North and South Ayrshire Councils), the Ayrshire Chamber of Commerce, the government housing agency, *Scottish Homes* (now Communities

---

[3] A typical definition of a 'new industrial district' as opposed to Marshall's notion of 'industrial district' can be found in Becattini (1990, p. 38), 'I define the industrial district as a socio-territorial entity which is characterized by the active presence of both a community of people and a population of firms in one naturally and historically bounded area. In the district, unlike in other environments, such as manufacturing towns, community and firms tend to merge'.

Scotland), the trades unions (through the Scottish Trades Union Congress – STUC) and *Scottish Enterprise*.   Since the mid 1990s the Forum has discussed the issues facing the county in the next half century and produced a consultative document, *A Vision for Ayrshire* (Ayrshire Economic Forum, 1998).   The idea was to include the whole community in a period of reflection and debate over the future development of the area.   This 'Challenge for the 21st Century' repeatedly stresses the need for partnership, co-operation and cohesion with recognition of 'our identity', 'our collective voice', 'our common cause', 'our Vision' and 'community'.   This approach to local economic strategic thinking and planning is now being encouraged across the country with local economic forums established for each area with a similar set of principal partners (see http://www.scotland.gov.uk/enterprise/localeconomicforums/).

Independent reviews of the local economy and local market bear out the usefulness of the partnership approach.   Thus *Scottish Homes* in Ayrshire promises that 'the majority of our [future] investment ... delivers against existing multi-agency commitments, with specific initiatives already identified and being prepared with partners' (*Scottish Homes South & West*, 1998, p. 29).   In a similar vein, South Ayrshire's strategy document '2020 Vision' is built on the idea of partnership and co-operation with the other agencies operating in the area, and beyond.   Joint campaigns in favour of infrastructure improvements (extension of the motorway system into the county) and in response to economic challenges (redundancies and closures) are strengthening progressively.

Within council areas, at a lower level of community, the Conservative government's programmes for *Priority Partnership Areas, Regeneration Areas* and *Smaller Urban Renewal Initiatives* continue to attract the support of a range of organizations, where agencies co-operate to reaping synergies supporting economic development.

At a higher level *Ayrshire Economic Forum* members co-operate in the competition for inward investment, EU funds and central government resources.   Such initiatives suggest a developing maturity in the partnerships first launched by the SDA and Strathclyde Regional Council in the 1970s.   The fact that they endure after considerable 'policy dislocation' (first the creation in 1991 of the local enterprise company network within *Scottish Enterprise*, the successor organization to the SDA, then local government reorganization in 1996) points to underlying strength in relationships between partners.   This basic trust supports continued co-operation.   But how do clusters, and in particular SE's cluster strategy, mesh in, if at all, with existing regeneration partnerships?   A review of developments in Ayrshire helps throw some light on this question.

Clusters at the County Level

In a development of the traditional industrial district concept the SE approach to clusters invites companies to establish partnerships building trust and shared visions for development.   Clusters are specialist networks that 'fuel innovation and generate synergies' enhancing international competitiveness (*Scottish Enterprise*, 1998, p. 1).   Clusters will encourage 'competition, co-operation and strong networks right across the community and between the public and private sectors'. The rhetoric, therefore, is similar to the regional economic development partnership model described above. Diagram 1, from SE suggests the existence or development of a coherent, inclusive partnership of all relevant development organizations. Resources, says SE, are to be focused on 'some of Scotland's most important value-creating clusters' – food, biotechnology, software, electronics, opto-electronics, and multimedia. Leaving to one side the geography of those sectors (most are under- or simply un-represented in Ayrshire) and the dependence on foreign-based multinationals, divergences seem to exist between the strategic approach as seen from SE headquarters and attempts at creating clusters locally.

Traditionally, textiles and engineering were two of Ayrshire's most important industry sectors in employment and sales. While both have a long history in the county they have also gone through major restructuring, with new firms and products appearing and well established enterprises being pushed out. Positive assessments of prospects and opportunities for the two sectors led to the creation of formal networking arrangements in the *Ayrshire Textile Group* (ATG) and the *Ayrshire Engineering Group* (AEG).

As elsewhere in Europe, but especially in the UK, the clothing and textile industry has been considered, since the late 1970s at least, to be a 'sunset industry' (Totterdill, 1992, p. 22). Structural adjustments in Scandinavia, Germany and Northern Italy to changing consumer tastes, competition from non-European countries and multi-national trading agreements prompted a reconsideration of the industry's future in Scotland, and in Ayrshire in particular. Research has suggested that successful adaptation requires:

- intensive market intelligence linked to continuous design innovation;
- high levels of technical expertise, particularly in production planning and problem solving;
- a highly versatile system of production;
- effective use of distribution networks. (Totterdill, 1992, p. 24).

As is well known, the Italian Emilia Romagna region has had great success in clothing and textiles and some have argued that a similar approach with a complex, multi-layered and decentralized model of partnership would also serve Ayrshire well (Totterdill, 1992, p. 47). While plans for Ayrshire supported such an initiative subsequent developments locally failed to restructure the industry, despite the creation of a partnership. Lack of understanding of the need to promote co-operation and trust was largely to blame (Danson and Whittam, 1998). Building trust, as is argued below, requires a degree of ownership by the constituent firms.

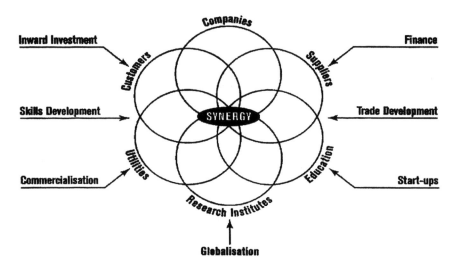

Figure 12.1   Scotland's Cluster Development Approach

Scottish Enterprise identified textiles as an industry where clustering should be encouraged. At the local level, Enterprise Ayrshire, (the local arm of SE) was instrumental in implementing the policy via the Ayrshire Textile Group (ATG). The textile industry in Ayrshire represents 25 per cent of Ayrshire manufacturing industry employing 9,000 people and exporting £60m per annum. In 1991 a partnership arrangement between Enterprise Ayrshire and the textile sector was established leading to the creation of the ATG in 1992. The commitment to partnership reflects the ethos of Enterprise Ayrshire's approach to local economic development seeking 'to develop and build relationships with companies rather than superimposing a structure onto a sector.' (ATG, 1994, p. 3). The ATG seemed initially to be successful in the 'delivery of information advice and programmes in response to proven needs ... with more than 60 per cent of local industry participating in some meaningful way' (ATG, 1994, p4). This led to product development, moving products up-market, increased diversification and marketing Ayrshire textiles as a group. Nevertheless the project failed to move forward to develop as a 'free-standing' resource centre recovering all its operating costs. Johnstone and McLachlan (1996) found textile companies reluctant to participate fully in the network making it difficult to realize economies of scale and scope, with, in addition, 'a low level of trust and the strongly adversarial nature of the [local] sector' (p. 755) a significant factor. Unwillingness to pool resources, concern over allocation of orders within the network, apprehension over co-operation, and communication failure helped explain lack of full commitment to the project. The need for training in networking protocol and processes, the key role of the facilitator, and the benefits of contractual agreements between members were identified as necessary elements if the local industry was to create a successful cluster. It seems that while ATG attempted to develop a partnership participating companies did not see the group as theirs.

Few companies, moreover, were willing to pay for ATG services. Where funds were provided for marketing, training and product development it was difficult to convince firms that they had to quantify 'outcomes'. A limited number of innovative enterprises did see the benefits of the partnership. Those companies tended to support joint seminars, the use of CAD/CAM, they recognized the potential of linkages with tourism, as well as technology transfer initiatives. In the local lace sector in particular, which produces some unique products and occupies significant niche markets, there was a strong reluctance to pass on skills, experience and expertise to younger recruits. This was bound to have serious consequences for the industry in Ayrshire.

Engineering Ayrshire (EA) was another clustering initiative in Ayrshire. Engineering in the 1990s was the largest manufacturing sector in the county with over 180 companies employing more than 14,000 people, a turnover of £1.9bn and generating £1.3bn in export sales (EA, 1997, p6). The original rationale for the establishment of the group was due to the perceived lack of sales outside the West of Scotland. To remedy this EA organized trade missions, launched a public procurement initiative and provided individual members with export sales and product/process improvement support. The EA group sought to widen its activities to capitalize on opportunities presented by changing market conditions. EA's structure gave grounds for optimism that the group could become self-financing, realising key objectives such as collective development of the skills base. Unlike the ATG the aims of EA state it should 'be an organization run by its members for the benefit of its members and the Ayrshire engineering and electronics sector' (EA, 1997, p. 9). Its strategic objectives place a responsibility on members to work in partnership with relevant organizations, such as Enterprise Ayrshire, local councils and trade associations. The group is managed by a steering committee of senior representatives from participating companies and Enterprise Ayrshire and meets monthly. A full meeting is held at least once a year. By placing obligations and responsibilities on the membership and by having an accountable framework for decision making it is hoped that the problems the ATG experienced can be avoided.

Conclusion

A cluster strategy can be an effective instrument in the regeneration of a regional economy. Scotland's trans-public sector partnerships generate synergies and mechanisms for resolving conflicts, a factor that has underpinned the appeal of the 'Scottish Partnership model' across the EU. Nevertheless, the lessons from this approach, as well as its theoretical underpinnings in the literature on industrial districts, have not been fully incorporated in existing strategies for industrial clusters. Trust and co-operation are essential ingredients if the advantages of innovation and networking are to be realized across sectors and if clusters are to be effective. In successful partnerships the long development of working relationships between organizations and individuals have been critical. But where networks have been dominated by multinational enterprises, or their structures and modus operandi have been imposed from the centre, significant difficulties have emerged, obscuring appreciation of the benefits that clusters can bring. Without some sense of ownership over *their* cluster and *their* network, the small and medium-sized enterprises that have most to gain from the promotion of the industrial districts idea will remain outside these promising partnerships.

References

Ashcroft, B. Love, J. (1996), 'Employment change and new firm formation in GB counties: 1981-89', in M. W. Danson (ed), *Small Firm Formation and Regional Economic Development*, London, Routledge.

Ashcroft, B. Love, J. and Schouller, J. (1987), *The Economic Effects of Inward Acquisition of Scottish Manufacturing Companies 1965-1980*, Industry Department for Scotland.

ATG (1994), *Business Plan*, Ayrshire Textile Group, 16 Nelson Street, Kilmarnock.

Ayrshire Economic Forum (1997), *Labour Market and Skills Trends in Ayrshire*, Ayrshire Economic Information Group, 17/19 Hill Street, Kilmarnock KA3 1HA.

Ayrshire Economic Forum (1997), *A Vision for Ayrshire*, 17/19 Hill Street, Kilmarnock KA3 1HA.

Beatty, C. and Fothergill, S. (1999), *Incapacity benefit and unemployment*, CRESR, Sheffield Hallam University.

Becattini, G. (1990), 'The Marshallian industrial district as a socio-economic.notion' in Pyke, F. Becatting, G. and Sengenberger, W. *Industrial Districts and Inter-firm Co-operation in Italy*, IILS, Geneva.

Boltho, A. and Holtham, G. (1992), The assessment: new approaches to economic growth. *Oxford Review of Economic Policy*, 8, 4, pp. 1-14.

Boyle, R. (1989) 'Partnership in practice: an assessment of public-private collaboration in urban regeneration, a case study of Glasgow Action', *Local Government Studies*, 15, pp. 17-28.

CEC (1993), *Growth, Competitiveness, Employment. The Challenges and Ways Forward into the 21st Century*, White Paper, Brussels.

Collins, C. and Lister, J. (1996), 'Hands up or heads up? Community work, democracy and the language of "partnership"', in I. Cooke and M. Shaw (eds) *Radical Community Work*, Moray House Institute of Education, Edinburgh.

Damer, S. (1990), *Glasgow: Going for a Song*, London, Lawrence and Wishart.

Danson, M. (1991), 'The Scottish economy: the development of underdevelopment?', *Planning Outlook*, 34, pp. 89-95.

Danson, M. Fairley, J. Lloyd, G. and Newlands, D. (1990), Scottish Enterprise: an evolving approach to integrating economic development in Scotland, in A. Brown and R. Parry (eds) *The Scottish Government Yearbook 1990*, Edinburgh University Press, Edinburgh.

Danson, M. Fairley, J. Lloyd, G. and Turok, I. (1997), *The Governance of European Structural Funds: The Experience of the Scottish Regional Partnerships*, Paper 10, Brussels, Scotland Europa.

Danson, M. and Whittam, G. (1998), Networks, innovation and industrial districts: the case of Scotland, in M. Steiner (ed) *From Agglomeration Economies to Innovative Clusters*, European Research in Regional Science, Pion, London.

DTI (1998), *Our Competitive Future - Building the Knowledge Driven Economy, Analysis*, 1998.

EA (1997), *Three Year Plan 1997-2000*, Engineering Ayrshire, 17/19 Hill Street, Kilmarnock KA3 1HA.

Freeman, C. (1994), 'The economics of technical change', *Cambridge Journal of Economics*, 5, pp. 463-514.

Grabher, G. and Stark, D. (1997), 'Organizing diversity: evolutionary theory, network analysis and postsocialism', *Regional Studies*, 31, 5, pp. 533-544.

Granovetter, M. (1985), 'Economic action and social structure: the problem of embeddedness', *American Journal of Sociology*, 91, 3, pp. 481-510.

Gregersen, B. and Johnson, B. (1997), 'Learning economies, innovation systems and European integration', *Regional Studies*, 31, 5, pp. 479-490.

Hodgson, G. (1988), *Economics and Institutions*, Cambridge, Polity.

Institute for Employment Research (2001), *Review of the Economy and Employment: Labour Market Assessment: Projections of Qualifications and Occupations – Volumes 1 and 2*, Coventr, IER, University of Warwick.

Johnstone, R. and McLachlan, A. (1996), 'The Ayrshire knitwear sector: from competition to collaboration', Paper to ISBA National Conference, Birmingham.

Lundvall, B.A. (ed) (1992), *National Systems of Innovation*, London, Pinter.

Lundvall, B.A. and Johnson, (1994), 'The learning economy', *Journal of Industry Studies*, 1, 2, pp. 23-42.

Malecki, E. and Nijkamp, P. (1988), 'Technology and regional development: some thoughts on policy', *Environment and Planning C*, 6, pp. 383-99.

Moore, C. and Booth, S. (1986), 'From comprehensive regeneration to privatization: the search for effective area strategies' in W. Lever and C. Moore (eds) *The City in Transition*, Clarendon Press, Oxford.

Morgan, K. (1997), 'The learning region: institutions, innovation and regional renewal', *Regional Studies*, 31, 5, pp. 491-503.

Randall, J. (1987), 'Scotland', in P. Damesick and P. Wood (eds) *Regional Problems, Problem Regions, and Public Policy in the United Kingdom*, Oxford, Clarendon Press.

Scott, M. (1992), 'A new theory of endogenous economic growth', *Oxford Review of Economic Policy*, 8, 4. pp. 29-42.

Scottish Council Development and Industry, (1997), *Exports Survey*, SCDI, Edinburgh.

Scottish Enterprise (1996), *Business Birth Rate Strategy: Update*, Scottish Enterprise, Glasgow.

Scottish Enterprise (1998), *The Clusters Approach*, Glasgow.

Scottish Enterprise (2000), *New Business Statistic and 'Inquiry 2000'*, www.newbusiness.org.uk.

Scottish Homes (1998), *Regional Plan 1998-2001*, Scottish Homes South & West, Paisley.

Slaven, A. (1975), *The Development of the West of Scotland: 1750-1960*, London, Routledge & Kegan Paul.

Storper, M. (1992), 'The limits to globalization: technology districts and international trade', *Economic Geography*, 68, pp. 60-93.

Storper, M. (1995), 'The resurgence of regional economies, ten years later: The region as a nexus of untraded interdependencies', *European Urban and Regional Studies*, 2, pp. 191-221.

Storper, M. and Scott, A.J. (1995), 'The wealth of regions', *Futures*, 27, pp. 505-526.

Totterdill, P. (1992), 'The textiles and clothing industry: a laboratory of industrial policy', in M. Geddes and J. Benington (eds) *Restructuring the Local Economy*, Longman: London. pp. 22-50.

van der Ploeg, F. and Tang P. (1992), 'The macroeconomics of growth: an international perspective', *Oxford Review of Economic Policy*, 8, 4, pp. 15-28.

Vaessen, P. and Keeble, D. (1995), 'Growth-oriented SMEs in unfavourable regional environments', *Regional Studies*, 29, pp. 489-506.

PART V

WHAT ROLE FOR FDI?

Chapter 13

Foreign Direct Investment in Scotland: The Silicon Glen Experience

Alistair Young

Filling the Gap

For much of the twentieth century, the Scottish economy has suffered from the decline of industries which had earlier been the source of its pre-eminence: shipbuilding, coal-mining, iron and steel manufacture. While the two world wars boosted demand for the outputs of these industries, they provided only a temporary respite: in the long run, decline was irreversible.

But there is a view of global economic development which sees the locational shift of industrial activities around the world as both inevitable and desirable. On this view, as companies in cheap labour locations in less developed countries learn to compete successfully in traditional industrial activities with those in more developed countries, the latter can shift resources into high-technology production instead. Following the second world war, industrial planners in Scotland came to see the electronics industry as a prime candidate to fill the gap left by the decline of traditional sectors. Scotland, it was claimed, had a plentiful supply of skilled technical and engineering labour which was available at rates lower than in many developed countries, both in Europe and the United States. The supply of this labour could be kept continually topped up by the local tertiary education system: Scottish universities and colleges had a long tradition of vocationally relevant engineering education.

A Faustian Bargain?

The attractions of the Scottish economy as a location for high-technology industry have been promoted by government agencies not only within the United Kingdom but also much further afield. In the years since the war, a major electronics industry has indeed developed within Scotland, much of it owned by multinational companies. To those who see globalization as a major force for encouraging a more effective use of the world's resources, the dominance of foreign firms in the Scottish electronics sector presents no problem. These companies may be seen as a vehicle for technology transfer, or for the passing-on of skills from foreign to local workers; or they may be welcomed as exemplars of new and more effective forms of industrial organization.

To other commentators, however, attempts to develop or restructure economies using foreign investment involve more of a Faustian bargain. A particularly strong statement of this view came from the 'New International Division of Labour' (NIDL) theorists of the early 1980s (Frobel, Heinrichs and Kreye, 1980). Although those authors were particularly concerned with the role in international production of less developed countries – the 'periphery' to the industrial centres of global capitalism – the kinds of argument which they put forward are frequently echoed in discussions of multinational activities in depressed regions of the industrial countries themselves. In the Scottish context, this discussion often finds expression in terms of the limitations of a 'branch plant economy'.

Briefly, such theories see the globalization of production activities as reflecting on the one hand the fragmentation of production caused by Taylorism, and on the other the reductions in costs of international transport and communications. Together, these factors allow the relocation of production activities to regions with low labour costs. On the most pessimistic view, these regions ultimately derive little economic benefit from the activities. Their role is to add value to the inputs which are then passed on to another geographical location for further processing. Hence there is little scope for backward linkages to stimulate suppliers within the local economy, nor yet for forward linkages to encourage the creation of local firms to make use of the output. Nor is there much genuine technology transfer: control and development of the technology remains firmly in the hands of headquarters, in the multinationals' home countries.

Although some value added is created locally, its extent is strictly limited. Most profits will flow back to headquarters. This process may be aided by transfer pricing. But even without this, the local tax regime is unlikely to take much of a slice of profits: favourable tax-subsidy policies, after all, would be among the instruments used by local planners to entice the multinationals in the first place. Since the other main enticement is a pool of cheap labour, it follows that the local wage component is unlikely to be very high either. The companies are alleged to ensure this by selective recruitment of young, often women, workers. In the strongest version of the NIDL, the local state apparatus is also seen as helping to keep wages low by legislating to prevent the efforts of workers to unionize to demand better pay and conditions. But in any case, market forces may well prevent any long-term improvement in the worker's share. If the pool of labour starts to dry up, and wages to rise, the companies will simply move to a hitherto unexploited location and begin the process anew.

Now, even as a characterization of the role of multinationals in less developed countries the NIDL hypothesis has been much criticized (see for example Wield and Rhodes, 1988; Henderson, 1989; Young, 1991). However, it provides a framework for thinking about the impact on local economies of the transnational division of labour, and if only for that reason it may be worth confronting it with evidence drawn from the Scottish electronics sector.

The Growth of the Scottish Electronics Sector

By the end of the war, Scotland already had an indigenous electronics sector, though not a very large one. In 1943, the British manufacturer Ferranti established a plant near Edinburgh, producing electronic equipment for the war effort. The defence sector continued to be a major market for UK electronics companies based in Scotland: not only for Ferranti, but also for Barr and Stroud, Marconi Instruments and Racal. It is true that by the early 1950s, the industry in Scotland was becoming more commercially oriented, but it was also increasingly dominated by American incomers, at first from the electromechanical business calculating machines industry: NCR, Honeywell, Burroughs and IBM. These were 'screwdriver' plants, set up to assemble components imported from America.

During the 1950s the business machine companies became involved in the development of electronic computers. At first these were based on valves; however, the discovery of the transistor in Bell Laboratories in 1947 permitted a switch to a much cheaper, more compact, more energy efficient and more reliable computer technology. Scotland became a location for the production of this crucial electronic component in 1960, when its first semiconductor plant was established in Glenrothes by the American company Hughes Aircraft Corporation: this was a diode assembly and test plant. As semiconductor technology developed, with the printing of integrated circuits and eventually microprocessors on a single chip, this sector continued to expand in Scotland, again through foreign direct investment. By the end of the 1960s Motorola, National Semiconductor and General Instrument had become established in Scotland; by the mid 1980s, other incomers had been added, including the defence electronics company Burr Brown and the Japanese company Nippon Electric (NEC). Another Japanese arrival, which was established in Livingston in 1984, was Shin-Etsu Handotai, a manufacturer of the silicon from which the chips were made.

As well as new investment, there was also disinvestment. General Instrument closed its Scottish operations in the mid-1980s. Burroughs merged with Sperry-Univac to form Unisys in 1986; soon after, its Cumbernauld plant was closed down. The workstation plant of Apollo Computers, similarly, was closed as part of the rationalization process following takeover by Hewlett Packard. Wang Laboratories was established in 1983 in close proximity to the University of Stirling, but closed in 1989. Timex, in Dundee, closed in 1993. Ferranti, arguably the founding firm in the Scottish electronics industry, went into receivership in the same year. In general, however, these more negative developments cannot be explained through any simplistic model of the global search for cheap labour. Often, as Clarke and Beaney (1993) note, they reflect the ways in which the corporations responded, or failed adequately to respond, to broad trends in technologies and markets. Thus the closure of Wang is to be explained by a failure on the part of the parent corporation to meet the challenge posed by the PC to its dedicated word processing systems. In the case of Timex, it is true that the company did seek cheaper labour locations during the 1970s; and there was a lengthy history of poor relations with the Dundee workforce. However, as Stephen Young has argued, the fundamental problem was 'slow and inadequate parent MNE response to technological change' (1984, p. 121).

On balance, however, the disappearance of electronics companies has been more than offset by new arrivals. Thus by the end of the 1990s, Scotland had acquired a production capability in a wide range of electronic products: optoelectronics, information systems manufacture, semiconductors, computer peripherals, telecommunications equipment, contract electronics manufacture and automotive electronics. The most obvious omission from the list is consumer electronics, which has always been relatively unimportant in Scotland, despite the arrival of Mitsubishi (video casette recorders and colour televisions) in the 1980s. Much of the output of the sector, about 80 per cent, was exported. According to *Locate in Scotland*, (2001, p. 8), at the beginning of the twenty-first century, Scotland produced 29 per cent of all personal computers made in Europe (and over 7 per cent of world PCs), 80 per cent of Europe's work stations, 29 per cent of Europe's Notebooks and 65 per cent of Europe's automated bank teller machines. Foreign owned companies continued to be dominant in the sector: 64 per cent of electronics employment was in multinational enterprises, 51 per cent being in American subsidiaries (*Scottish Enterprise*, 1999).

Surveys of the multinationals have revealed a number of motivations for locating in Scotland. The availability of a cheap pool of labour, stressed by the NIDL theorists, is certainly a factor. However, by contrast with assembly plants in less-developed countries, the required pool of labour for the Scottish electronics industry has had to include not only semi-skilled workers but graduate managers and engineers. Thus as Henderson notes (1989, p. 128), the fact that electronics engineers in Scotland could be hired at only 60 per cent of the cost of their counterparts in the United States was a significant inducement. Henderson also points out that the tendency for social welfare costs to be borne by the state through general taxation, rather than (as often in other European countries) through employee benefits, greatly reduces the cost of labour to the employer.

Labour costs, however, have been only one of several locational attractions. An important 'market-based' factor was the desire to gain access to the European market. As an English-speaking country, Scotland provides a convenient location for production for export to continental Europe. After 1973, when the UK became part of the European Economic Community and hence found itself within the 17 per cent Common External Tariff, the attractions of a UK production location were reinforced. In the case of semiconductors, for example, companies could produce the bulk of their value added in their Scottish wafer-fabrication plants, assemble and test them in East Asian plants, and re-import them for sale within the European community (Henderson, 1989 p. 126).

Other factors leading to a Scottish location include such infrastructural advantages as good communications by road and air; the presence of nearby research-oriented universities; and even the attractions to managers of proximity to St Andrews golfing facilities (McCalman, 1988). Not least among the incentives, however, was the support of government agencies, though there is some dispute over just how important this factor really was. While McCalman's respondents listed government financial aid as easily the most important locational determinant (op. cit. pp. 123-4), Henderson (1989, pp. 131-2) considered it 'very much a secondary determinant', and noted that inward electronics investment continued in the 1980s even although regional development grants had been reduced to half their late 1960s level.

Government Support for the Electronics Sector

Government support takes many forms. In recent years, it has chiefly been provided through the agency, *Locate in Scotland*, which was established in 1982.[1] This agency provided information on alternative sites and local economic conditions (labour and skill availability and cost, transport and communications infrastructure, the local supply network). It also provided information on investment and training incentives.

Much of 'Silicon Glen' is in fact located within 'Assisted Areas', as defined under Section 7 of the 1982 Industrial Development Act: 85 per cent of employees of overseas-owned electronics firms were based in these areas in 1992 (Jackson and Patel, 1996, p. 17 n. 2). Indeed, substantial numbers of the incoming companies are located within 'development areas', which qualify for the highest levels of assistance. This assistance takes the form of discretionary grants payable towards the cost of equipment purchase, land purchase, site preparation and buildings. Recent trends in this form of assistance are shown in Table 13.1.

Table 13.1   Regional Selective Assistance (RSA)

	1994-95	1995-96	1996-97
Manufacturing:			
(a) value of offers accepted (£m)	94.4	87.4	146.8
(b) new and safeguarded jobs	12,100	11,270	15,470
(a)/(b) (£)	7,802	7,755	9,489
Electronics and optical:			
(a) value of offers accepted (£m)	50.9	34.6	91.4
(b) new and safeguarded jobs	5,480	3,910	8,950
(a)/(b) (£)	9,288	8,849	10,212

Source: Scottish Office, *Economic Bulletin,* March 1998.

As Table 13.1 shows, the costs to public funds of creating or sustaining jobs in electronics is rather greater than for manufacturing as a whole; this is, perhaps, not surprising, given the capital-intensive character of this sector. It is also clear from the table that electronics takes a substantial share of the funding available under RSA: over half, in two of the three years detailed in the table.

Official policy, then, encouraged heavy investment in the expansion of the Scottish electronics industry, particularly through the attraction of foreign-owned enterprise. This has helped to build up one of the largest concentrations of electronics companies outside the United States. But how worthwhile was the effort? Clearly, this may be assessed according to a number of criteria. In the following parts we consider the effects on employment, wages and conditions, the linkage effects on local

---

[1] *Locate in Scotland* was joined in October 2001 with *Scottish Trade International*, to form *Scottish Development International*.

suppliers, future prospects for the sector, and (even more speculatively) the possible broader implications of an active and diverse electronics sector for the development of a 'knowledge economy' in Scotland.

Employment in Silicon Glen

In making any appraisal of the impact of the electronics industry on the Scottish economy, an immediate problem concerns the basic information available. Official statistics, unfortunately, are not collected specifically for the electronics sector, which is scattered over a number of categories. Over time, however, a number of attempts have been made to devise estimates of value added and employment in the electronics sector. It should be noted, however, that figures over time are not always strictly comparable, since the basis for the estimates may have changed.

Table 13.2   Employment Trends in Scottish Electronics, 1945-99

	1945	1959	1971	1983	1989	1995
Employment in the Scottish electronics industry	3,000	7,400	37,200	42,500	40,768*	41,100

Note: *Adjusted to allow for non-electronics employment (see note to Table 13.3 below).

Source: 1945-83, Firn and Roberts (1984); 1989, Jackson and Patel (1996); 1995, Scottish Office.

A breakdown of the employment within the sector, taken from the Scottish Input-Output Tables for 1989, is as follows:

During the 1990s, the electronics industry was the most significant growth area in Scottish manufacturing. An indicator of this is given by the Index of Industrial Production: while for manufacturing as a whole this increased from 100 to 116 between 1990 and 1996, the increase for 'electrical and instrument engineering' over the same period was from 100 to 237 (*Scottish Economic Bulletin*, March 1998). As will be clear from the figures in Table 13.1, the growth in output is to be explained by productivity increases rather than by employment expansion. Indeed, the Scottish Office (1998) notes that between 1992 and 1995 average annual productivity growth in electronics was 23 per cent, as compared with only ten per cent in manufacturing as a whole.

Table 13.3   Employment and Value Added in Scottish Electronics, 1989

		Employment ('000)	Employment (% share)	Value-added (% share)
44	Office machinery and computer equipment	10.7	24	45
47	Industrial electrical equipment	2.7	6	4
48	Telecommunication related equipment	14.4	32	23
49	Electronic components	9.4	21	17
50	Electronic consumer goods	1.6	4	5
57	Instrument engineering	6.0	13	8
	Total	44.8	100	100

Source: Jackson and Patel (1996, p. 19). The total differs from the 1989 figure in the previous table because it includes some non-electronic employment, thought to be especially significant in groups 47 and 57.

Thus the recent impressive expansion of the electronics sector has not been matched by any corresponding increase in employment. Moreover, a comparison of the figures in Table 13.2 with those in Table 13.1 is a little disturbing; it suggests that a substantial proportion (latterly, over a fifth) of the jobs within electronics are maintained only by substantial subsidies from the public purse. True, in addition to the 41,000 thought to be directly employed in Scottish electronics in the late 1990s, *Locate in Scotland* claims that 29,900 were employed in support activities. Even when such activities are taken into account, however, the electronics sector is clearly making only a modest contribution towards filling the gaping hole in employment opportunities left by the decline in traditional industries. To give some indication of the magnitudes involved, it may be noted that in the Census of Population employment categories 2, 3 and 4, covering the manufacturing sectors, the total number of employees in Scotland in 1951 was 771,000; by 1993, despite the contribution of the electronics sector, this figure had fallen to 365,000 (calculated from Lee, 1995, Table 1.3).

What about the quality of employment opportunities in the electronics companies? As noted earlier, these are by no means confined to semiskilled or unskilled workers. In the four largest semiconductor plants in the mid-1980s, Henderson estimated that 26 per cent of all employees were engineering or technical staff. However, a recent study by McNicholl (2000) suggests that Scottish electronics firms have a somewhat less highly-skilled profile than electronics firms in the rest of the UK. Thus 'professional and intermediate' skill groups amounted to 35 per cent of the workforce in Scottish plants, but to 42 per cent for plants in the rest of the UK; the corresponding figures for 'partly skilled' workers were 31 per cent and 23 per cent respectively.

On the question of union protection for workers, with some exceptions (IBM and National Semiconductor in Greenock, for example), the companies have preferred to locate in areas with no strong tradition of union militancy, such as the Scottish New

Towns. Henderson also notes that they have tended to recruit women workers, particularly younger ones; and have sought, generally with success, to avoid recognition of unions. Unions may, of course, have had a positive indirect effect on wages and conditions in these plants, to the extent that companies have tried to pre-empt demands for union representation by offering reasonable terms to their workers. A study by Brand et al (2000, pp. 348-9), using data for 1994, notes that wages in the Scottish electronics sector in that year were marginally below the regional average.

Linkage Effects

One of the principal questions asked about the effectiveness of the electronics strategy concerns the breadth and depth of the linkages which it has made with the rest of the economy. This is of interest for several reasons.

First, there is the question of permanence. As we have seen, among the reservations often expressed about the globalization of production is that multinational firms have no strong locational loyalties; as cheaper or more profitable locations become available in other parts of the world, they will shift to them. However, firms which have well-established links with local suppliers and subcontractors are less likely to find it profitable to go off elsewhere than those whose sources of supply are already more distant.

Then there is the question of how far the presence of large multinational firms has succeeded in stimulating local production. A.O. Hirschman (1958), one of the first authors to emphasize inter-firm linkages as a key aspect of development strategy, saw them as a way of economising on one of the most scarce of resources, decision-making or entrepreneurial skills. On this view, the arrival of new companies in an intermediate industry, such as electronics, should induce decision-making either to supply its inputs or to utilize its outputs.

Linkages also give rise to technology transfer. The development of local sources of supply to high technology enterprises requires that indigenous workers or companies engage in a learning process to acquire new skills or technological competencies. There is a 'public goods' or 'externalities' element to the acquisition of such skills, since they may subsequently be used for the benefit of enterprises other than those for which they were initially acquired. The presence of such externalities, in principle at any rate, provides some justification for the use of tax revenues to encourage the high technology industries to establish themselves in the first instance.

So how well embedded are the multinational electronics within the Scottish economy? This question has been asked in several studies of the sector; the results, so far, have generally been somewhat discouraging. Thus an early study undertaken for the *Scottish Development Agency* (1980) showed that only 12 per cent of electronics components and 30 per cent of subcontracting work were sourced in Scotland; by contrast, the corresponding figures for purchases from the rest of the UK were 46 per cent and 54 per cent. These figures could be regarded as measuring opportunities for Scottish companies. However, later studies do not suggest that they have been very successful in doing so.

Table 13.4   Linkage Effects in Scottish Electronics

Scottish sourced material inputs by value (% share)

SDA (1979)	19
Macalman (1984-5)	15
SDA (1986)	12
SDA (1988)	15
Jackson and Patel (1989)	17
SDA (1990)	12
Turok (1991)	12

Scottish sourced material and service inputs by value (% share)

Brand et al (1994)	28

Note: Dates refer to time of surveys, not of publication. The results of Jackson and Patel are from the Scottish Input-Output Tables; Brand et al have updated the Input-Output tables using Census of Production and their own sample survey data. Other figures are also from sample surveys.

In 1984 and 1985, McCalman (1988) undertook a survey of the 30 enterprises in the foreign-owned electronics sector, in which he sought to test the hypothesis that linkages might develop over time. This might conceivably happen for two reasons: first, because MNEs might need time to identify a network of local suppliers on whose quality they could rely; second, because as the local branches of the MNEs themselves matured they might be given more autonomy from head office, with the implication that they might use this autonomy to develop local sources of supply. Unfortunately, the evidence from the firms surveyed did not support this chain of reasoning. Those firms that arrived in Scotland prior to 1970 were no more likely to purchase from local sources than those that came later. This was so even though the majority of MNEs had devolved purchasing responsibility on their local subsidiaries. The managers in the foreign-owned firms made a number of suggestions to explain the low level of local sourcing: Scottish companies, it was alleged, lacked the necessary technical competence, they were mostly too small to be able to cope with the size of orders which the MNEs wanted to place, and they were in any case unwilling to commit themselves to dependency on multinational customers. In a wide range of performance characteristics relating to quality and flexibility, Scottish companies were ranked as inferior to suppliers elsewhere in the UK, or in the USA or Europe.

Similar results to those of McCalman were recorded by a series of *Scottish Development Agency* surveys (SDA, 1986, 1988, 1990), and by a postal survey conducted by Ivan Turok in 1991, as shown in Table 13.4 above. Despite differences in the sample of firms used, the figures have a narrow range, and suggest that only 12-15 per cent of material inputs by value were being sourced from within Scotland during this period. Turok's study shows that foreign firms within the sector were much more likely to import components than were UK firms. He also shows that some of the largest 'local' suppliers were themselves foreign owned, rather than indigenous to Scotland. This was true in the area of Printed Circuit Board assembly; although this is a sector where there have been relatively low barriers to entry as well as advantages from having close links between buyers and suppliers, Scottish companies have failed to make many inroads into the market. Local indigenous suppliers seem to have been most successful in the simplest and bulkier manufacturing processes, including in particular packaging materials. Although almost half by value of plastics were sourced locally they were often produced by subsidiaries of companies located elsewhere in the UK, or abroad.

Turok's findings at the level of the individual companies are supported at the more aggregate level by comparisons of gross output and gross value added in the sector over time (Tables 5 and 6). Between 1983 and 1989 gross output at current prices, which includes the value of input purchases, grew from £1,573 million to £4,939 million. Gross value added, which is net of input purchases, grew only from £616 million to £1,197 million. The ratio of value added to gross output thus fell, from 39.1 per cent to 24.2 per cent. Within the foreign-owned sector the corresponding fall was from 34.6 per cent to 20.2 per cent, local input content being even smaller.

Table 13.5   Ratio of Gross Value Added to Gross Output, 1983-89

	1983	1984	1985	1986	1987	1988	1989
All firms	0.39	0.35	0.34	0.33	0.31	0.29	0.24
Foreign	0.35	0.31	0.30	0.28	0.27	0.25	0.20

Source: From Turok (1993), Tables 1.2 and 1.3.

Turok concludes that the linkages between foreign electronics companies and local suppliers corresponded more closely to what he describes as a 'dependent' rather than a 'developmental' model. The key distinction here is that the latter model offers prospects of 'self-sustaining growth through cumulative expansion of the industrial cluster', whereas in the former the suppliers are 'vulnerable to external forces and corporate decisions'.

Turok's results gain some support from an analysis by Jackson and Patel (1996) of the 1989 Scottish Input-Output Tables, although the ratio of purchases from Scottish suppliers, at 17 per cent, is slightly higher than the figures quoted previously. Of all industries and services, electronics has much the lowest proportion of Scottish purchases (the average ratio is 50 per cent). However, Turok's approach has been criticized by McCann (1997). At the aggregate level, as McCann notes, the observation of a fall in the propensity to source locally could be explained not so much by an increasing tendency of existing firms to go elsewhere for suppliers, as by the arrival of newcomers on the scene. The latter are likely to have a lower propensity to make local purchases, though this may improve over time. Furthermore, the trend over the 1980s for previously vertically integrated companies to contract out activities which had been carried on in-house would automatically lead to an apparent fall in the value-added- to-gross-output ratio. Finally, by comparing 1979 and 1989 input-output tables for Scotland, McCann claims to find evidence of strengthening rather than weakening local linkages.

In response, Turok (1997) has argued that McCann's emphasis on the differing local sourcing propensities of new and established firms would be unlikely to explain the sharp fall in the value added ratio, since the bulk of the expansion in gross output during the 1980s was in fact from established, rather than from new firms. Scepticism regarding this aspect of McCann's argument might perhaps be further strengthened by the evidence, earlier quoted, from McCalman's survey, which casts doubt on whether new arrivals do in fact have significantly different purchasing patterns from more etablished firms. On the question of the general trend towards outsourcing, Turok points out that this would only inflate gross output artificially if the outsourcing had been within Scotland, which for the most part it was not. Further, he criticizes McCann's comparison of 1979 and 1989 data on the grounds that this focuses on employment and income multipliers, rather than on the more relevant output multipliers. Turok also cites more recent evidence from a *Scottish Enterprise* survey of the 16 largest foreign non-semiconductor electronics companies (Table 13.6): this shows that the ratio of purchases was about 21 per cent in 1994-95 (the inclusion of semiconductor plants would probably have reduced this figure). This again seems to support a more pessimistic interpretation of the scope for development of local linkages.

The figures in Table 13.4 from the study by Brand et al (2000) give a broader picture of the linkage effect by including local sourcing of services as well as of materials. However, the same authors have also compared the foreign owned Scottish electronics sector with UK owned firms, with other industrial sectors, and with firms in Wales and the West Midlands. In general, foreign owned Scottish electronics does poorly in these comparisons. The figure quoted in Table 13.4 is less than for UK-owned electronics firms in Scotland (37 per cent), and is the second lowest among the foreign owned firms in the five industrial sectors studied. It is higher than for foreign owned electronics firms in Wales (25 per cent), but lower than the corresponding figure for the West Midlands (35 per cent).

Table 13.6   Local Purchases by 16 Largest Foreign-Owned Electronics Companies

1991-92	26.8%
1992-93	20.6%
1993-94	18.8%
1994-95	20.6%

Source: *Scottish Enterprise*, Annual Surveys (figures exclude semiconductor plants and also purchases of electronic components and services and inter-company trading).

Table 13.7   Distribution of Purchases, 1994-95

Scottish Spend %	
Storage media	0
Display units	1
PCBA	43
Higher level assembly	78
Cables	45
Power supply	12
Plastics/rubber	38
Input devices	12
PCB	16
Sheet metal/presswork	58
Printed material	69
Packaging	93
Castings/machined parts	40

Source: Scottish Enterprise.

Future Prospects for the Electronics Industry in Scotland

On the positive side, Jackson and Patel offer some evidence of a likely upward shift in local sourcing in the area of display units. As data from *Scottish Enterprise* show, this is a very large item of import expenditure by local electronics companies (Table 13.7). However, the agreement of Tatung Electronics of Taiwan to establish the Chung Hwa Picture Tubes cathode ray manufacturing plant in Lanarkshire has led to a substantial rise in local sourcing, even if not from an indigenous Scottish company. This development was actively sought by *Scottish Enterprise* to fill a gap which had been identified in the PC supply chain.

However, it is becoming clear that the emphasis in Scottish Executive policy is shifting away from the attraction of inward investment. A recent statement of the new emphasis is given in the Scottish Executive document, *Scotland: a Global Connections Strategy,* published in October 2001 (Scottish Executive, 2001b). At various points in that document, the growing limitations of the inward investment policy are acknowledged:

> Overall, US foreign direct investment is declining relative to the EU ... Other European locations now occupy the low cost niche aimed at securing untrammelled access to EU markets (Ministerial Foreword).

> ... a substantial proportion of overseas investment continues to be in lower skilled manufacturing and assembly operations which would today go elsewhere and which are proving increasingly vulnerable to closure and relocation (p. 13).

Developments in the sector during 2001, with large numbers of employment cuts being announced in electronics, added force to those comments. An estimate for the first seven months of 2001 put the total job loss at over 6,000 (The Herald, August 1, 2001). Job losses were particularly severe as a consequence of the closure of the Motorola factory in Bathgate where over 3,000 jobs disappeared as well as the downsizing and, ultimately, closure of the NEC plant at Livingston, involving 1,260 jobs. A further 860 jobs were lost at the Compaq plant in Erskine. In the first two cases the companies explained job losses as being due to a slump in demand for the product: mobile phones at Motorola and DRAM semiconductors at NEC. Compaq's job cuts were a consequence of global restructuring involving the shifting of production to outside suppliers: in this case, to a Taiwanese company operating in the Czech Republic. This last example seems particularly pertinent to the concerns expressed by the Scottish Executive as quoted above. Elsewhere in that document, the implications of EU enlargement are also mentioned, both as increasing the attractiveness for FDI of new and low-cost members, and also as reducing Scotland's share of support from structural funds (ibid, p. 9).

In response to these challenges to traditional strategy, the Executive makes a number of proposals for change. In particular, it emphasises the need to take advantage of the opportunities allegedly being made available through the development of the 'knowledge economy'.

Towards a Knowledge Economy?

The concept of the 'knowledge economy' creates some problems of definition. A recent OECD definition includes both high technology industries, such as electronics and software, and sectors with a highly skilled work force, such as finance and education. The presumption is that as economies develop to more advanced levels these knowledge-intensive activities will increase as a proportion of GDP. For present purposes, the relevant question concerns how to demonstrate the existence and importance of mutually-reinforcing linkages between electronics and such activities.

This is a much trickier task than estimating traditional inter-industry linkages, so what follows is necessarily impressionistic. However, it seems reasonable to expect to find 'knowledge-linkage' effects in the following areas:

- between universities and technical colleges on one hand, and the electronics sector on the other. Scotland, compared with the rest of the UK, has a greater proportion of students in tertiary education, as well as a strong tradition of engineering education and education-industry linkages. As we have seen, this was a factor encouraging foreign electronics firms to locate in Scotland; however, the interaction with such companies can also have highly beneficial feedback effects on higher education (see for example Connor, Wylie and Young, 1986);
- between electronics and the software sector. As with the electronics industry itself, the software sector has been expanding rapidly in Scotland (Table 13.8). According to *Locate in Scotland*, the software community is growing at 15 per cent annually, and currently consists of 19,600 specialists in companies, universities and research institutes. As far as software companies themselves are

concerned, there are over 550. There are obvious linkages with the electronics companies: for example, 2,000 advanced software engineers were employed by IBM, Compaq, Sun and Hewlett-Packard in 1999.

Table 13.8 Software in Scotland

- 15 per cent annual growth rate of software workforce
- 19,600 software specialists in companies, universities, R&D institutes
- software market of £700 million per year
- growth rate of software market 3x European growth rate
- National Software Strategy predicts extra 30,000 software jobs in next 10 years
- More mathematical and computing science graduates per head than America, Japan or Western Europe
- 550 software companies
- Softnet development centres in 10 Scottish towns, covering 60 firms

Source: Locate in Scotland, 1999.

• between electronics and design. While the multinational electronics companies were considered by Henderson to make only a modest contribution to design, which tended not to be seen as a Scottish function, he also notes that this was an area in which small indigenous firms were more active, particularly spinoff companies from Edinburgh University (1989, pp. 151-2). A recent example of new investment in design is the formation of the Alba Centre in Livingston, for the design of system-on-chip semiconductor devices.

The example of the Alba Centre, however is two-edged. A report in the Financial Times (7 October 1999) noted that one of the principal participants in the Centre, the American company Cadence Design Systems (now known as Tality) had scaled back its recruitment plans on the grounds of a worldwide shortage of engineers and managers with relevant experience. While this may have posed a short-term problem for the centre, it may perhaps be seen more positively as evidence of the global importance of the type of activities which the centre was set up to encourage. It also gives force to the emphasis in the recent Scottish Executive documents to the need to make Scotland an attractive location for highly skilled people to live in (Scottish Executive, 2001a, p. 12).

Conclusion

If FDI in the Scottish electronics industry had been seen as a complete remedy for the restructuring problems caused by the decline of traditional heavy engineering industry, it would obviously have been a disappointment. It is insufficiently labour intensive to fill the employment gap; and its relative failure to develop backward linkages also limits its usefulness as a source of economic stimulation.

On the positive side, it has been possible to persuade some of the world's leading-edge electronics companies to take root in Scotland and to set up technologically sophisticated processes. This has involved the training of generations of Scottish workers in advanced engineering and management skills; it has also provided one of the few remaining sources of Scottish manufacturing export growth. It would be fair to say that for the last 40 years at least this sector has made a respectable contribution to the host economy. It is also fair to note that the importance of this contribution will probably become steadily less in the future, as has been rightly recognized in recent strategy papers from the Scottish Executive. As yet, it is too early to judge whether the new strategies proposed in those documents, based on the 'knowledge economy', will serve as an effective replacement.

References

Brand, S. Hill, S. and Munday, M. (2000), 'Assessing the impacts of foreign manufacturing on regional economies: the cases of Wales, Scotland and the West Midlands', *Regional Studies*, (34), (4), pp. 343-355.

Clarke, T. and Beaney, P. (1993), 'Between autonomy and dependence: corporate strategy, plant status, and local agglomeration in the Scottish electronics industry', *Environment and Planning A*, (25), pp. 213-252.

Connor, A. I. Wylie, J. A. and Young, A. (1986), 'Academic-Industry liaison in the UK: Economic perspectives', *Higher Education* 15: pp. 407-420

Firn, J.R. and Roberts, D. (1984), 'High technology industries', in Young, S. and Hood, N. *Industry, Policy and the Scottish Economy*, Edinburgh, Edinburgh University Press, pp. 317-325.

Frobel, F. Heinrichs, J. and Kreye, O. (1980), *The New International Division of Labour*, Cambridge, Cambridge University Press.

Henderson, J. (1989), *The Globalisation of High Technology Production*, London, Routledge.

Hirschman, A.O. (1958), *The Strategy of Economic Development*, New Haven, Yale University Press.

Jackson, N. and Patel, D. (1996), 'Local sourcing by the electronics industry in Scotland', *Scottish Economic Bulletin* (52), pp. 17-29.

Lee, C.H. (1995), *Scotland and the United Kingdom*, Manchester, Manchester University Press.

Locate in Scotland, *General Information Brochure*, August 2001.

McCalman, J. (1988), *The Electronics Industry in Britain: Coping with Change*, London, Routledge.

McCann, P. (1997), 'How deeply embedded is Silicon Glen? A cautionary note', *Regional Studies*, (31), (7), pp. 697-704.

McNicholl, I. (2000), 'Industrial demand for skilled labour: Scotland and the UK', *Quarterly Economic Commentary*, XXV/3.

Scottish Development Agency (1980), *The Scottish Electronics Subcontracting and Component Supply Industries*, Survey undertaken by Makrotest, Glasgow, SDA.

Scottish Development Agency (1986, 1988, 1990), *Scottish electronics companies surveys*, Glasgow, SDA.

Scottish Enterprise (1999), *Key Facts about the Scottish Economy.*

Scottish Executive (2001a), *A Smart, Successful Scotland.*

Scottish Executive (2001b), *Scotland: a Global Connections Strategy.*

Scottish Office (1998), *The Electronics Industry in Scotland.*

Turok, I. (1993), 'Inward investment and local linkages: how deeply embedded is Silicon Glen?', *Regional Studies*, (27) (5), pp. 401-417.

Turok, I. (1997), 'Linkages in the Scottish electronics industry: further evidence', *Regional Studies* (31) (7), pp. 705-711.

Wield, D. and Rhodes, E. (1988), 'Divisions of labour or labour divided?' in Crow, B. Thorp, M. *et al*, *Survival and Change in the Third World*, Oxford, Polity Press/ Basil Blackwell.

Young, A. (1991), 'The semiconductor industry: its contribution to third world development', *Science, Technology and Development*, 9 (1&2), pp. 199-208.

Young, S. (1984), 'The foreign owned manufacturing sector', in Young, S. and Hood, N. *Industry, Policy and the Scottish Economy*, Edinburgh, Edinburgh University Press, pp. 93-127.

Chapter 14

FDI in Upper Silesia – Experience and Lessons

Adam Drobniak

Introduction

In many post-communist countries foreign direct investment (FDI) is often perceived as a dynamic factor in the process of economic development. FDI is not only a form of financial transfer but also transmits knowledge, know-how and raises business standards. FDI contributes to building co-operative relationships with domestic enterprises and to creating technological spill-over effects (Wojnicka, 1999, pp. 42-43). In the last decade, as economic transformation began, Central and East European countries (CEECs) have started to play an active role in attracting foreign capital. But the FDI inflow to those countries, in comparison to the overall sum of world foreign investment, was small and amounted to merely 4.6 per cent, although in the period 1990-97 it increased 30-fold (Przybylska, 2001, p. 22). In the case of Poland FDI plays a key role both in privatization and new business formation. At the regional level we are particularly interested in the impact of FDI on building a durable basis for development and increasing competitiveness. We are also well aware of course that regions compete to attract inward investment. A successfully competitive region is one where the level of human knowledge, understood as an ability to be ahead of needs and to be innovative, creates structural advantages (Markowski, 1996, p. 7).

The main factor determining a high level of any region's competitiveness is innovativeness. An innovative system is perceived as a group of mutually inclusive, connected regional institutions that support the process of economic innovation. FDI that increases human knowledge is highly favourable for regional development and strengthening its competitive position. Competitive advantage is based on a complex configuration of resources, regional synergy and increased role of intellectual factors (Markowski, 1996, p. 5). FDI plays the role of the exogenous impulse that stimulates endogenous potential.

A different situation occurs when foreign business activity is separated from regional (domestic) enterprises and institutions. Then co-operative networks are not created. Only the most basic abilities of human capital (like relatively unskilled manual work) are utilized, the imported technology remains quite separate from the local technology base. In such regions competitive advantage is based on existing resources or their simple configurations, and regions are unable to create and absorb innovations. The advantage gained by foreign investors is usually a result of spatial differences in the prices of production factors like labour. The main risk associated with that kind of foreign investment is that as the differences in factor prices diminishes the movement of investors to other regions becomes more likely. Foreign investors are not likely to be involved in the long run.

There is of course vigorous competition among regions in trying to win foreign investors (Bojar, 2001, p. 12). In addition, it is normally expected that FDI will flow to less developed areas when products' life cycles are mature. From the regional viewpoint skilful selection of FDI should pay attention to the foreign investor's sector, the volume of capital involved and the likely impact of investment on the regional economy, its human capital, as well as its ability to play an active role in the region's techno-productive systems.

The Silesia Region

In Poland, Silesia along with Mazowieckie (the Warsaw region) and Wielkopolskie (the Poznań region) lead in attracting foreign investors. During the 1990s Silesia attracted around ten per cent (more than US$3bn) of the foreign capital inflow to Poland. Within Silesia the best performing districts (*powiaty*) in attracting investment (measured by foreign companies per 10,000 inhabitants) were Katowice, Bielsko-Biała, Tychy and Gliwice. The largest foreign investors tended to be attracted to the same locations but also to Częstochowa, Sosnowiec and Żywiec.

According to the Polish Agency for Foreign Investments (PAIZ) in the year 2000 more than 260 firms with investments exceeding US$1m were active in Poland. Figure 14.1 shows foreign investment in Poland by region in first half of 2000.

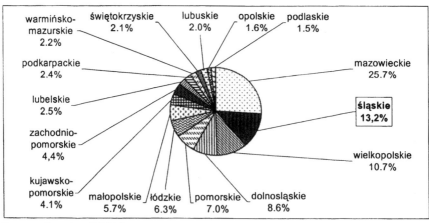

Source: Based on PAIZ data.

Figure 14.1 Foreign Investment in Poland by Region (First Half of 2000)

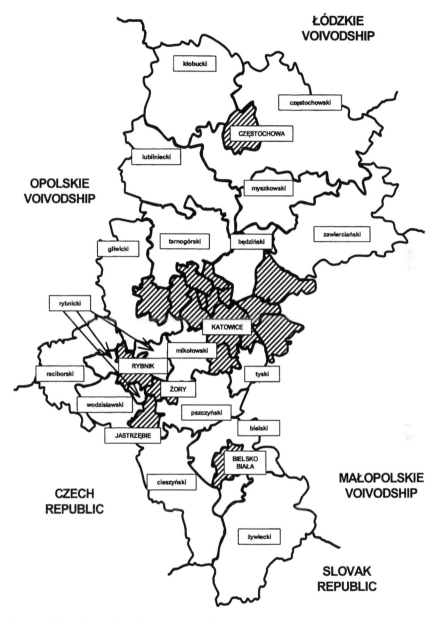

Figure 14.2   Silesia's Administrative Structure
(The Region's 36 Districts, *powiaty*)

PAIZ estimates that, over 1990-2000 foreign investment in Poland totalled just over US$49.4bn, an average of just US$1,300 per capita. In the best year (2000) the inflow was US$10.6bn. Nearly half of the foreign investment in 2000 was connected with privatization and the second half was mainly in 'greenfield' projects. The foreign capital inflow fell back considerably in the year 2001.

The Silesian voivodship has 36 territorial units at the National Unit for Territorial Statistics (NUTS) 4 level. Of those, 17 are rural and 19 cities or towns with district status. Generally, towns and cities have much higher population density, 1,604 people per km^2 in 1999 as compared to 190 people per km^2 in rural areas. They also had, unsurprisingly, most of the fixed assets of enterprises, more than two-thirds of the regional total and most investment, 79 per cent of the total in 1999. Katowice district accounts for the greatest investment activity (US$1,604 per capita), Gliwice, just to the west, comes next (US$1,411 per capita), then Tychy in the south (US$1,036 per capita). The most neglected areas in respect of investment are the more peripheral districts of Kłobucki (US$46 per capita), Wodzisławski (US$61 per capita) and Rybnicki (US$61 per capita).

The central part of the voivodship is dominated by the Katowice Agglomeration. Its great significance is shown by the fact that it has nearly 44 per cent of the voivodship population, 45 per cent of all enterprises and almost 57 per cent of fixed assets. Investment outlays per inhabitant amount to US$631, that is, 147 per cent of the voivodship average (US$430). To put it in another way, the central area receives 64 per cent of all the investment outlays in Silesia. The Appendix gathers basic data on the Katowice Agglomeration.

Beyond the central part of the voivodship three sub-regions can be identified. The northern sub-region around Częstochowa has a low level of infrastructure services, low potential for small and medium sized enterprises (SMEs) and agricultural predominates in the sub-region's economy structure. The southern sub-region around Bielsko-Biała, has a relatively diversified economic structure, great potential for SME activity, tourism development connected with the Beskidy mountains but a poorly developed transport infrastructure. The western sub-region around Rybnik, which along with the nearby towns Żory and Jastrzębie forms the Rybnik Agglomeration has a low level of development in business services, an economic mono-structure dominated by coal mining, and high unemployment. In Żory the unemployment rate reached 27.8 per cent in June 2001 while the average unemployment rate in Silesia was 14.2 per cent.

The specificity of the sub-regions briefly described above was one of the main reasons for establishing the Katowice Special Economic Zone (KSSE) in 1996. KSSE is unusual among special zones in that it is dispersed with a total area of 827 ha in four sub-zones: Gliwicka with 336.4 ha; Sosnowiecko-Dąbrowska with 221.8 ha; Tyska with 143.5 ha and Jastrzębsko-Żorska with 125.3 ha. In 2000 another two areas in Częstochowa (the area around the steel works) and Bielsko-Biała (around the FIAT plant) were added to the KSSE. The idea behind the special economic zone is to support and accelerate restructuring and to create new jobs in Silesia. The most important investment incentives are profits and real estate tax relief. Investors get 100 per cent profits tax relief for 10 years and 50 per cent tax relief for the remainder of time until the end of the zone's existence in August

2016. KSSE is an exceptional zone in that it has attracted 40 per cent (US$970m) of all investment outlays across all the Polish zones. Up to April 2000 nearly 60 businesses had taken advantage of the incentives on offer in the KSSE, *promising* the creation of 14,000 jobs. However in practice the zone experiment had relatively little impact on employment with only about 8,000 new jobs, and that, it is worth emphasising, is less then 0.4 per cent of all workplaces in Silesia.

The special economic zone contains several sub-zones and some have had huge difficulty in attracting investment. For example the Jastrzębsko-Żorska sub-zone in the western part of the region attracted merely 3.6 per cent of the total KSSE investment. This lower investment attractiveness is due to the smaller market, the predominance of unskilled labour, the local economy's mono-structure (dominated by coal). Other sub-zones have been more popular and have attracted big investors including GM Opel, FIAT, ISUZU Motor, Ekocem, Mahle, Delphi Automotive Systems, ROCA Radiadores and Granges AB.

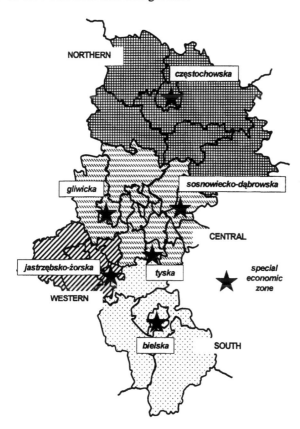

Figure 14.3 Sub-Regions and the Katowice Special Economic Zone (KSSE)

The KSSE did play some part in promoting restructuring. The zone attracted about US$1bn of FDI as greenfield investment, that is, 30 per cent of total FDI in Silesia. Of course any objective evaluation of the KSSE requires a thorough cost-benefit type analysis and some direct benefits plainly exist including capital inflow, new jobs, and creation of co-operation networks between foreign investors and Polish companies. On the down side special economic zones have caused considerable controversy in EU accession negotiations.

Foreign Direct Investment in the Region

Investment in Total

On a per capita basis *total* investment (domestic and foreign) in Silesia in 1999 amounted to US$430. As might be expected the greatest investment was in the region's cities and towns (US$575) rather than in rural areas where investment outlays reached only US$218. The evidence shows that investment concentrates on districts in the central part of the voivodship (see Figure 14.4). The best performing districts were Katowice (US$1,603), Gliwice (US$1,411), Mikołowski district (US$1,113) and Tychy (US$1,034) where investment per inhabitant exceeded the average several times over. The downside is that the concentration of investment in the central part of Silesia deepens sub-regional differentiation and further marginalizes northern and western areas of the voivodship.

The second group comprises areas that are characterized by a relatively good position with regard to investment per inhabitant. Those areas with outlays in a range +/- 50 per cent of the average (US$215 to US$640) are focused on northern and the north-eastern parts of Silesia (Częstochowa, Częstochowski, Zawierciański, Tarnogórski, Będzieński counties, Dąbrowa Górnicza, Sosnowiec, Chorzów). Two areas in the southern sub- region (Żywicki district and Bielsko-Biała) are also in this group. The case of Bielsko-Biała is especially interesting because its investment, at US$630 per person, shows that the city performs well and nearly makes it into the best performing group. Two other towns from the western sub-region (Rybnik with US$493 and Jastrzębie with about US$400) also get into to our second group.

The third group contains the areas with low inward investment, well below average (less than 50 per cent of the average, that is below US$215 per person). These are mainly rural districts and peripheral areas around the centre or sub-centres of the region.

While analyzing the investment structure in Silesia one should note that the total amount of investment in 1999 was US$2.1bn, from which little more than one half (54 per cent or US$1,127m) came from the private sector. Foreign investment (worth US$413m) was nearly 20 per cent of the region's total investment and accounted for 37 per cent of private sector investment.

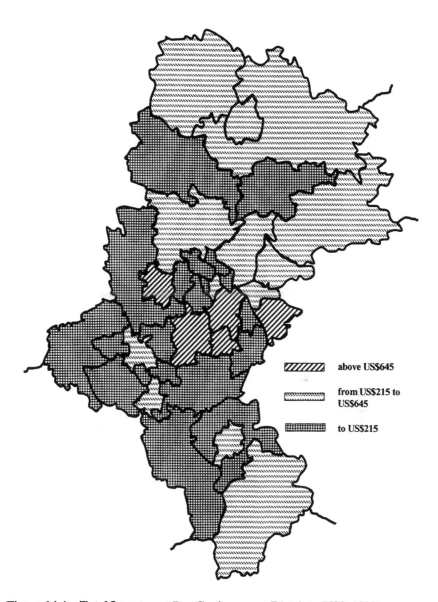

Figure 14.4   Total Investment Per Capita across Districts (US$, 1999)
(Average across all Districts, US$430)

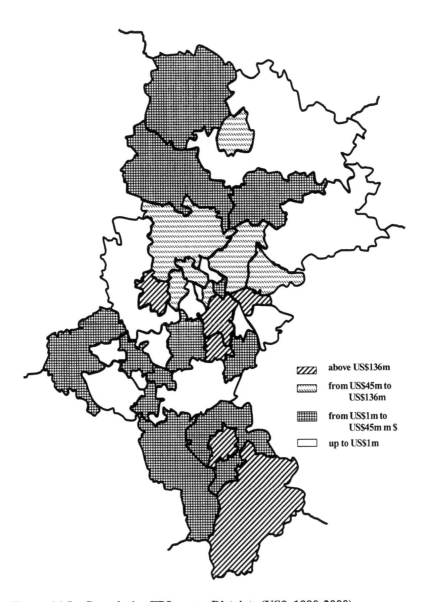

Figure 14.5   Cumulative FDI across Districts (US$, 1990-2000)
(Average per District across Entire Region, US$90.8)

Table 14.1   Total Investment in Silesia in 1999

Investment Outlays by Ownership Sector	(Zl m)	(US$ m)	(%)
TOTAL	8,631.3	2,094.9	100.0
Public sector	3,984.8	967.2	46.2
Private sector of which:	4,646.5	1,127.8	53.8
• private domestic ownership	1,752.3	425.3	20.3
• foreign ownership (including partnerships with dominance of foreign capital)	1,703.2	413.4	19.7
Total investment per capita		430.0	

Source: Regional Statistical Office in Katowice.

FDI in Silesia 1990-2000

It is clear that FDI is important to the region especially in the private sector where it accounts for more then one third of total business investment. FDI strengthens regional economic growth, not only complementing domestic investment but playing a distinctive role. Up to the end of 2000 Silesia had attracted FDI of US$3,230m, that is, nearly US$91m per district. The spatial allocation of cumulative FDI across districts is shown in Figure 14.5. The best performing districts were Bielsko-Biała (US$1,298m, mainly FIAT), Katowice (US$546m, due to Nationale Nederlanden Group in banking and insurance), Gliwice (US$380m, mainly GM Opel in vehicles), Tychy (US$259m, mainly investment of ISUZU Motors and Delphi Automotive Systems in vehicles and by South African Breweries in the drinks branch); Sosnowiec (US$130m, FIAT and Timken Company in vehicles); Żywiecki county (US$195m, PepsiCo and Heineken in food and drink and Alpha company in packaging). A detailed list of main foreign investors is shown in Table 14.2 below.

A less well performing group, in which FDI ranges from US$40 to US$100m, is focused on districts that surround or belong to Katowice Agglomeration. They include Tarnogórski   (US$70.3m, European Bank of Reconstruction and Development in banking), Bytom (US$61.0m with main investors OBI Systemzentrale and Metro AG, both in retail/wholesale trade, Statoil in petrol stations), Będziński (US$57.0m, Metro AG in retail and wholesale trade and Saint Gobain in chemicals), Dąbrowa Górnicza (US$48.9m, Air Liquide in technical gases, Metro AG and Saint Gobain as well as ABB in electrical machinery), Zabrze (US$46.4m, Metro AG and BP Amoco in petrol stations), Chorzów (US$46.0m with Alstom in transport machinery and equipment as well as Statoil, Częstochowa (US$45.1m, with British Oxygen Corporation BOC in chemicals, Metro AG BP Amoco and Statoil.

Analysis of FDI with regard to the value of investment per capita confirms, in principle, the conclusions drawn from the analysis of FDI in relation to districts, that is, urban areas do better. Cumulative FDI per capita reached US$672 by 2000 with a spread in absolute terms of US$7,200 between best and worst performers, a huge spatial diversity. However since investments of less than US$1m are excluded in these PAIZ data this diversity is likely to be exaggerated. Figure14.6 shows that the concentration of areas with the highest level of FDI per capita is clearly in the central and southern part of the region.

FDI per capita in cities like Katowice (US$1,603), Gliwice (US$1,411) and Żywiec district (US$1,291) is several times higher than the region's average (US$672). A real record holder is Bielsko-Biała, where FDI per inhabitant exceeds the region's average by more than ten times, mainly as a result of investment made by FIAT. The second group consists of the areas where FDI per inhabitant fits the range from 50 per cent to 100 per cent of the region's average and included Sosnowiec (US$537), Tarnogórski (US$488), Będziński (US$385), Chorzów (US$380) and Dąbrowa Górnicza (US$374).

Most FDI in Silesia comes from EU countries (more than 88 per cent) with greatest capital inflow from Italy (40 per cent) mainly because of FIAT investment in vehicles and from Germany (20 per cent) due to GM's Opel plant in Gliwice. Supermarket investment is also important. FDI from the Netherlands (17 per cent) was the third source, its focus on banking and insurance (Nationale Nederlanden Group) and in drinks (Heineken). Some 80 per cent of foreign investment in Silesia originates from these three countries. Silesia's attractiveness for foreign (mainly EU) capital is obvious – it has provided an early spatial re-location opportunity enabling firms to relocate parts of production chains into Central European countries (cheap labour, low risk factors). Of course the planned EU enlargement helps too. However investors' decisions may be temporary – the base for future investment in post Soviet countries.

As stated earlier foreign capital enters Poland either via share purchase or in completely new investment, either brown- or green-field. The major investments in the first group, via privatizations, include Fiat in Bielsko-Biała and Tychy, Heineken in Żywiec, Nationale Nederlanden Group's purchase of the biggest Silesian bank – Bank Śląski, Alstom's purchase of a tram plant in Chorzów, Air Liquide's purchase of a technical gases plant, Henkel's purchase of chemical plant, British Oxygen Corporation's purchase of metal products manufactures in Częstochowa, South African Breweries' purchase of the Tychy brewery. Such privatizations involved in the first instance companies with good economic positions and development capabilities. Firms taken over by foreign capital benefited, and in most cases continue to benefit, mainly by strengthening their competitiveness, such as modernization, new product development and improved access to new markets. However, foreign investors in Silesia have avoided investing in firms from traditional sectors such as steel – and of course the restructuring of those sectors is the main problem in Silesian economic development.

Table 14.2　Principal Investors in Districts with Highest Cumulative FDI, 1990-2000

District	Investor	Branch	Investment (US$m)
	FIAT	Vehicles	1,143.0
	EBRD	Banking	73.6
Bielsko-Biała (US$1,298m)	Simest	Capital investment	31.0
	European Renaissance Capital L.P.	Retail and wholesale trade	20.0
	Metro AG		20.0
Katowice (US$546m)	International Nederlanden Group	Banking and insurance	324.0
	IAEG – WIBERA	Construction	66.0
	MAHLE	Vehicles	35.2
	ROCA Radiadores	Manufacture of non-metallic products	22.3
	Granges AB	Capital investment	18.9
	INPRO	Manufacturing basic metals	17.5
	Manuli Rubber Industries	Vehicles and manufacturing of rubber and plastic products	16.4
Gliwice (US$38m)	GM Opel AG	Vehicles	360.0
	Saint Gobain	Chemicals	n.a.
Tychy (US$259m)	ISUZU	Vehicles	55.6
	Delphi Automotive Systems	Vehicles	34.7
		Drinks	25.0
	South African Breweries	Vehicles	n.a.
	FIAT		
Sosnowiec (US$130m)	FIAT	Vehicles	81.3
	Timken Company	Manufacturing of machinery	12.5
	Metro AG	Retail and wholesale trade	17.5
Żywiecki	Heineken	Drinks	180.7
	Alpla	Packaging	14.0

Source:　Own calculations.

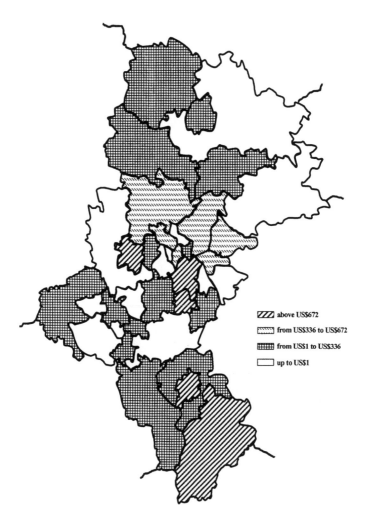

Figure 14.6   Cumulative FDI Per Capita across Districts (1990-2000, US$)
         (Average for Entire Region, US$672)

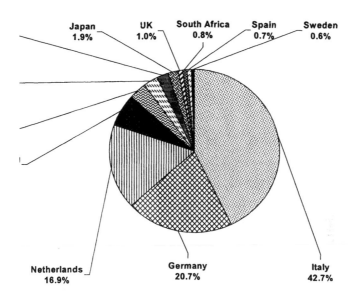

Figure 14.7 Sources of FDI

As for greenfield investments the most important have included the GM Opel plant in Gliwice, the ISUZU plant in Tychy, the Delphi Automotive Systems plant, the MAHLE plant, the ROCA Radiadores, the Granges AB plant (all these investments are in the KSSE), the Alpla plant (packaging in Żywiec) and the retail and wholesale trade network of Metro AG, Tesco, Géant, IKEA. The greenfield investments in the vehicle industry have led to a change in the region's economic structure. Vehicle and trailers in 1999 accounted for nearly 16 per cent of sold production, exceeding the share of basic metals (14.8 per cent), and not much below the hard coal share (18.8 per cent).

We conclude this review of FDI in Silesia with an analysis of the number of foreign-owned enterprises (3,453 firms in the region at the end of 1999). They were concentrated in the central part of the voivodship and in Częstochowa, Cieszyński and Tarnogórski districts. The situation of the southern sub-region also looks good, particularly in Bielsko-Biała. By far the worst performing parts are in the northern (Zawierciański, Kłobucki, Lubliniecki, Myszkowski districts) and western sub-regions (Rybnicki county, Jastrzębie, Żory). Foreign firms' locations are shown in Table 14.3 and in Figure 14.8.

Table 14.3   Foreign Firms in the Overall Number of Companies

Highest Participation			Lowest Participation		
County	No. of Firms	%	County	No. of Firms	%
Katowice	687	19.9	Powiat rybnicki	11	0.3
Bielsko-Biała	342	9.9	Powiat zawierciański	15	0.4
Gliwice	252	7.3	Powiat tyski	19	0.6
Częstochowa	251	7.3	Powiat kłobucki	21	0.6
Powiat cieszyński	184	5.3	Powiat lubliniecki	22	0.6
Powiat tarnogórski	143	4.1	Powiat myszkowski	22	0.6
Tychy	130	3.8	Piekary Śląskie	24	0.7
Zabrze	112	3.2	Jastrzębie	26	0.8
Bytom	104	3.0	Świętochłowice	28	0.8
Sosnowiec	102	3.0	Żory	32	0.9

Source: Own calculations using regional Statistical Office data.

The spatial diversity of FDI and location of foreign companies confirms that investors prefer the areas within the central part of the voivodship. But in some cases, a relatively low level of FDI may by associated with a relatively large number of foreign firms. For example this appears to be the case in Tarnogórski and Cieszyński districts. The reverse is also possible as in Żywiecki where a high level of FDI is associated with a small number of foreign firms. Such relations indicate that in some areas foreign capital is concentrated in one or a few foreign firms. FDI in Cieszyński and Tarnogórski is dispersed among a relatively high number of small and medium sized enterprises (SMEs). We should bear in mind however that investments below US$1m are not included in the PAIZ data we have utilized here.

Summing up, comparing the number of foreign enterprises to FDI per district (or per capita) we can draw a few conclusions on FDI types. In the first type high FDI per capita is associated with a small number of enterprises per district meaning that FDI is concentrated (on one or a few large firms). In the second type a high level of FDI per capita is associated with a large number of foreign firms, suggesting that FDI is diversified. This looks good for local development but the downside is that local firms can become strongly dependent on foreign enterprises with little endogenous development. In the third form a low level of FDI per capita is associated with a large number of firms suggesting that FDI is mainly an SME phenomenon. This is the most desirable form of FDI because the foreign sector enters into local production systems but does not play a dominant role.

We turn finally to the factors determining foreign investors branch preferences. In 1999 foreign firms were in trade and repairs (40.8 per cent), manufacturing (27.5 per cent), real estate (9.3 per cent), construction (7.9 per cent), transport, storage and communication (6.5 per cent), hotels and restaurants (2.4 per cent). Trade, manufacturing and real estate accounted for nearly 70 per cent of the total number of all foreign companies. Most FDI has been in trade with the huge number of super and hypermarkets its best showcase. The strong competitive position of large scale foreign retailers unfavourably affects relations with suppliers (sometimes payments can be delayed), employees (in practice trade unions rarely exist in supermarkets) and small retailers (under severe pressure).

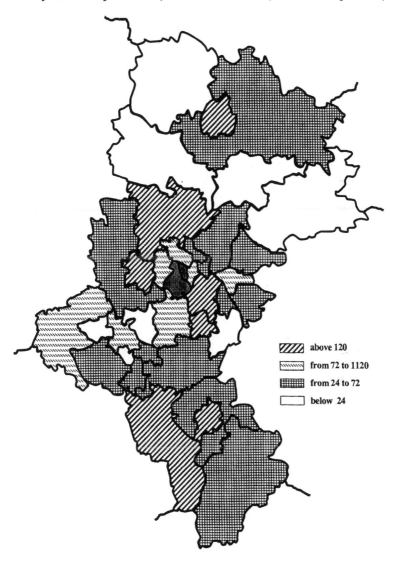

Figure 14.8   Foreign Companies Spatial Location (numbers)

Conclusions

Foreign capital searches the regions for the best rate of return. Silesia along with Mazowieckie and Wielkopolskie is considered to be that kind of region. FDI is attracted to Silesia because of its large market and local purchasing power, its human capital with its skills and experience, its industrial culture and favourable location on Central European transport corridors.

Within the region, investors mainly prefer the central area: it has the greatest socio-economic potential. The KSSE's success in attracting FDI is less significant than the factors discussed above. Indeed the evidence shows that more disadvantaged areas (such as Jastrzębsko-Żorska) with lower market potential and higher unemployment were not in any substantial way helped by the existence of the KSSE.

Without a doubt FDI has made an important contribution to changing the region's economic profile from domination by coal and steel to a more diversified structure including vehicles (FIAT, GM Opel, ISUZU). But those structural changes have established another industry (motor vehicles) that is also exceptionally sensitive to economic conditions and was badly hit in the recent downturn. All the major producers with assembly operations in Poland announced large scale redundancies in 2001-02. Although substantial methodological problems get in the way of establishing the precise extent of smaller FDI projects it is of interest that Cieszyński district, where the total value of FDI is relatively low, the total number of foreign enterprises is far above the average per district within the region. Clearly in Cieszyński foreign capital is allocated mainly among SMEs and that kind of foreign investment is definitely better for regional development than concentrated FDI.

An important issue for regional transition is the impact of FDI on the ability to create local production systems. In many cases, especially for regions with a heavy industry heritage, such systems may help create regional innovation networks that will be at the basis of a new competitive economic structure. Local production systems are usually defined as durable relations among enterprises flowing from co-operation between large firms and SMEs as partnerships or sub-contractors (Jewtuchowicz, 2000, p. 148).

Analysis of the impact of FDI in Silesia suggests that foreign investors have not created durable relations with Polish SMEs in the last decade. Services, or sub-assembly products or components contracts are limited only to low technology items. Indeed there are numerous examples of R&D departments shutting down in companies privatized with foreign capital. Foreign enterprises build co-operation and sub-contract relationships only reluctantly. Most components are imported. This kind of behaviour is not favourable to building and promoting the local production system or to forging strong foreign investor identity with and commitment to the region.

FDI's impact on employment in the region is hard to establish. No reliable statistics exist on job creation. The only data relate to the promise to create some 14,000 new jobs in the KSSE but the outcome looks to have been considerably fewer (around 8,000). Foreign-owned enterprises are said to have generated 80,000 jobs but that is only four per cent of the total employment in Silesia. Lack of data relating to jobs, the impact of privatization and stiffer competition, complicates rational analysis of this issue. Despite the great amount of FDI in Silesia the unemployment rate increased from 12 per cent in 2000 to more then 14 per cent at the beginning of the year 2002.

Appendix

Katowice Agglomeration

City / Town	Population ('000)	Enterprises absolute numbers	Fixed Assets Per Capita (US$)	Investment Per Capita (US$)
Katowice Agglomeration	2,119.3	163,169.0	7,491.2	631
Voivodship	4,865.5	371,243.0	5,784.3	430
Katowice share (%)	43.6	45	130	147

Source: Regional Statistical Office in Katowice.

City / Town	Population ('000)	Enterprises Absolute Numbers	Fixed Assets	Investment
			Per Capita (US$)	
Katowice	343.2	38,692	11,031.3	1,607.4
Bytom	203.8	13,153	3,592.1	115.9
Chorzów	121.2	8,747	5,495.4	225.8
Dąbrowa Górnicza	130.9	11,554	16,108.7	370.7
Gliwice	210.8	18,047	9,450.2	1,415.0
Jaworzno	97.5	6,120	11,086.2	978.8
Mysłowice	79.3	5,964	6,030.6	85.0
Piekary Śląskie	65.7	3,214	1,800.2	173.8
Ruda Śląska	156.8	8,156	7,490.4	164.8
Siemianowice	76.9	5,136	3,737.5	155.8
Sosnowiec	242.3	21,635	4,610.2	267.1
Świętochłowice	59.0	3,383	4,638.9	89.2
Tychy	132.7	11,780	7,673.3	1,036.2
Zabrze	199.2	11,588	5,235.8	151.7

Beyond the central part of the voivodship three sub-regions can be identified. The northern sub-region around Częstochowa has a low level of infrastructure services, low potential for small and medium sized enterprises (SMEs) and agricultural predominates in the sub-region's economy structure. The southern sub-region around Bielsko-Biała, has a relatively diversified economic structure, great potential of SMEs, development of tourism activities connected with the Beskidy mountains but a poorly developed transport infrastructure. The western sub-region around Rybnik, which along with the nearby towns Żory and Jastrzębie forms the Rybnik Agglomeration. This area has a low level of development in business services, an economic mono-structure dominated by coal mining, and high unemployment. In Żory the unemployment rate reached 27.8 per cent in June 2001 while the average unemployment rate in Silesia was 14.2 per cent (June 2001).

References

Błaszczyk, A. (2000), 'Integracja w obliczu zagrożenia' (Integration in the face of threats), *Rzeczpospolita*, August 24.

Bojar, E. (2001), *Bezpośrednie inwestycje zagraniczne w obszarach słabo rozwiniętych*, (FDI in weakly developed regions) Wydawnictwo Naukowe PWN, Warszawa.

Jewtuchowicz, A. ed, (2000), *Strategiczne problemy rozwoju miast i regionów* (Strategic problems in developing towns and regions), Zakład Ekonomiki Regionalnej i Ochrony Środowiska, Uniwersytet Łódzki, Łódź, 2000.

Markowski, T. (1996), 'Od konkurencji zasobów do konkurencji regionów' in *Regionalne i lokalne uwarunkowania i czynniki restrukturyzacji gospodarki Polski. Wzrost konkurencyjności regionów*, Friedrich Ebert Stiftung, Łódź.

Przybylska, K. (2001), *Determinanty zagranicznych inwestycji bezpośrednich w teorii ekonomii* (Determinants of FDI in economic theory),Wydawnictwo Akademii Ekonomicznej w Krakowie, Kraków.

Rosati, D. (2001), 'Inwestycje zagraniczne wspierają złotego' (Foreign investment holds the złoty up), *Rzeczpospolita*, July 7.

Stępniak, A. Ed, *Swobodny przepływ pracowników w kontekście wejścia Polski do Unii Europejskiej*, (The free flow of labour in the context of Poland's EU entry), Urząd Komitetu Integracji Europejskiej, Warszawa.

Trębski, K. (2001), 'Polowanie na kapitał', (The chase for capital), *Wprost*, July 29.

Regionalne i lokalne uwarunkowania i czynniki restrukturyzacji gospodarki Polski. Wzrost konkurencyjności regionów, (Regional and local conditions and factors in restructuring the Polish economy: the growth of regional competitiveness), Friedrich Ebert Stiftung, Łódź, 1996.

Rocznik Statystyczny Województwa Śląskiego 2000, Urząd Statystyczny w Katowicach, 2000.

Statystyka Powiatów Województwa Śląskiego 2000, Urząd Statystyczny w Katowicach, 2000.

Raport o stanie przemysłu w roku 1999, (Report on the state of industry in 1999), Ministerstwo Gospodarki, Warszawa, 2000.

Raport o stanie handlu wewnętrznego w roku 1999, (Report on the state of internal trade), Ministerstwo Gospodarki, Warszawa, 2000.

Wojnicka, E. (1999), *Bezpośrednie inwestycje zagraniczne w procesie prywatyzacji gospodarki polskiej*, (FDI in privatizing the Polish economy), Instytut Badań nad Gospodarką Rynkową, Gdańsk, 1999.

www.paiz.gov.pl – Polish Agency for Foreign Investment.

www.silesia-region.pl – Silesia Regional Government – Office for Investor Service.

Index

Printed and bound by CPI Group (UK) Ltd, Croydon, CR0 4YY

22/10/2024

01777626-0005